One Hundred and One Botanists

One Hundred and

One Botanists

DUANE ISELY

Iowa State University Press / Ames

To botany graduate students who read essays in the form

©1994 Iowa State University Press, Ames, Iowa 50014
All rights reserved

Authorization to photocopy items for internal or personal use, or the internal or personal use of specific clients, is granted by Iowa State University Press, provided that the base fee of $.10 per copy is paid directly to the Copyright Clearance Center, 27 Congress Street, Salem, MA 01970. For those organizations that have been granted a photocopy license by CCC, a separate system of payments has been arranged. The fee code for users of the Transactional Reporting Service is 0-8138-2498-2/94 $.10.

⊛ Printed on acid-free paper in the United States of America

First edition, 1994

Library of Congress Cataloging-in-Publication Data
Isely, Duane.
   One hundred and one botanists / Duane Isely.—1st ed.
     p.   cm.
   Includes bibliographical references (p.   ) and index.
   ISBN 0-8138-2498-2 (acid-free paper)
   1. Botanists—Biography. I. Title. II. Title: 101 botanists.
QK26.I75   1994
581'.092'2—dc20
[B]
                                                                                                                             94-28380

Unless otherwise noted, the portraits of botanists are published courtesy of the Hunt Institute for Botanical Documentation, Carnegie Mellon University, Pittsburgh, Pa.

of a weekly column and asked, *"Why don't you put these together as a book?"*

CONTENTS

Preface, *ix*

Aristotle, *3*

Theophrastus, *6*

Dioscorides, *10*

Hildegard of Bingen, *14*

Otto Brunfels, *17*

Luco Ghini, *20*

Hieronymus Boch, *23*

Piendra Mattioli, *26*

Valerius Cordus, *29*

Rembert Dodoens, *32*

Konrad Gesner, *35*

Andrea Cesalpino, *39*

Clusius, *43*

John Gerard, *46*

Gaspard Bauhin, *49*

Jan Baptista Van Helmont, *53*

John Ray, *56*

Anton Leeuwenhoek, *60*

Robert Hooke, *65*

Nehemiah Grew, *68*

Joseph Pitton de Tournefort, *71*

Rudolph Camerarius, *74*

Stephen Hales, *77*

John and William Bartram, *80*

George-Louis (Comte de) Buffon, *82*

Carl Linnaeus, *86*

Victor von Haller, *94*

Michel Adanson, *97*

Johann Hedwig, *101*

Jan Ingenhousz, *104*

Joseph Priestley, *107*

Sir Joseph Banks, *110*

Jean Baptiste de Monet de Lamarck, *113*

Antoine-Laurent de Jussieu, *118*

Thomas Knight, *121*

Christiaan Persoon, *124*

Johann Moldenhawer, *127*

Nicholas de Saussure, *129*

Robert Brown, *132*

Frederick Pursh, *136*

René Dutrochet, *139*

Amos Eaton, *143*

Augustin Pyramus de Candolle, *145*

William Jackson Hooker, *148*

Thomas Nuttall, *151*

Elias Fries, *154*

John Henslow, *157*

John Torrey, *160*

George Bentham, *163*

Adolphe-Theodore Brongniart, *167*

Miles Berkeley, *170*

Jacob Schleiden, *173*

Hugo von Mohl, *176*

Alphonse de Candolle, *178*

Alvan Wentworth Chapman, *181*

Charles Darwin, *184*

George Engelmann, *188*

Asa Gray, *191*

Joseph Dalton Hooker, *196*

Carl von Nägeli, *200*

Johann Gregor Mendel, *203*

Nathanael Pringsheim, *207*

Wilhelm Hofmeister, *210*

Heinrich de Bary, *213*

Julius von Sachs, *216*

Julius Brefeld, *220*

August Eichler, *223*

Phillippe Van Tieghem, *225*

Johannes Warming, *227*

Heinrich Engler, *230*

Eduard Strasburger, *223*

Charles Bessey, *237*

William Pfeffer, *241*

Hugo de Vries, *244*

Luther Burbank, *247*

John Coulter, *251*

Marcus Jones, *254*

Gottlieb Haberlandt, *258*

Dukinfield Henry Scott, *261*

Frederick Orpen Bower, *264*

Karl von Goebel, *267*

Andreas Franz Wilhelm Schimper, *271*

Liberty Hyde Bailey, *274*

Roland Thaxter, *278*

Nathaniel Lord Britton, *280*

George Washington Carver, *285*

Hugh Neville Dixon, *291*

Charles Deam, *293*

Albert Spear Hitchcock, *296*

Daniel MacDougal, *299*

Mary Agnes Chase, *303*

Henry Chandler Cowles, *306*

Henry Horatio Dixon, *308*

John Kunkel Small, *311*

Arthur Tansley, *315*

Richard Willstätter, *318*

Merritt Fernald, *321*

Frederick Clements, *326*

Agnes Robertson Arber, *331*

Henry Gleason, *334*

Winona Hazel Welch, *340*

Index of Botanists, 345

Incidental Index, 347

PREFACE

🐚 BOTANISTS WE WILL NOT KNOW

The first botanists, heroes of their time, whose names have come to us are Theophrastus (ca. 370 B.C.-285 B.C.) and Dioscorides (ca. 20-70 A.D.). They are given their moment on the marquee in following pages. But certainly there are botanists before them, nameless to us and all of history, innovative persons whose findings were used by people in successive generations for untold years. This supposition is supported by the extensive knowledge and lore about plants that existed before recorded history and that was inherited by pioneer botanists of name.

Dentation suggests that most subsistence for umpteen years came from plant materials, seeds, roots, and tubers. The eating of flesh was probably secondary. Before agriculture, it was food gathering. Perhaps nearly everyone, men and women alike, knew the useful plants, where to find them, how to recognize them, and when and how to gather them. In addition to food, the plant resource extended to drug, oil seed, medical, and ritual plants. Before and beyond all of this seemingly universal know-what, -where, and -how, there are certainly some individuals who went beyond established tradition, who explored and found new ways of using old plants, and discovered new plants that could be added to the inventory. (Indeed, modern exploration and science has not yielded any more really new food plants.) These individuals were the professionals who developed and sequentially added to the reservoir of knowledge; they are those botanists whom we will never know.

🌿 BOTANISTS WE KNOW: ABOUT THE ESSAYS HERE

This book consists of essays about botanists, students of plants, arranged chronologically from Aristotle (384-322 B.C.) to the near present with George Washington Carver (1864-1943) and ending in the 1990s with Winona Welch. Included are mostly those who significantly contributed, over an epoch of more than 2,000 years, to making botany (plant science) what it is today.

It is not too hard to learn about the professional life and contributions of named botanists of the past if you know how to get around in a good library. It is often difficult, however, to obtain information about the life of botanists as persons, for example, married or not? Family? Personal eccentricities? Except for a few for whom there is a full biography, the botanist is treated as a disembodied spirit and biographic reference is all business. For example, "von Hohenheim wrote such-and-such books, discovered chlorophyll in *Rhizopus nigricans* (bread mold), and after studying ferns looked for alternation of generations in muskrats." They don't tell that he had an ear shot off by a jealous husband and that he hated cats.

These botanists, indeed not disembodied spirits, were individuals who lived busy and usually impassioned lives. Their existence included suffering as well as the joys of living and accomplishment. Most of our heroes were "good" or "nice" people, but a few were not. If the hawg is a hawg is a hawg, it is presented as such, warts and all.

There seems to be a difference between botanists and those in the humanities. Take musicians, for example. In my home library I have a couple of shelves on music. There one easily knows what kind of persons were Mozart, Beethoven, and Liszt, and of their nonprofessional adventures. And of the great novelists, the same. But botanists, at least to the public eye, have mostly been souls of propriety. Boring by comparison? I don't think so. Certainly I hope you will not find either these accounts or their botany wearisome.

In references there is much about honors, society memberships, complimentary degrees, etc., that come to our stellar botanists. These are mostly not listed here. They can be taken for granted if the person belongs to the elite.

PREFACE   xi

The research has included almost no primary investigation, i.e., searching out the "papers" (letters to and from others, unpublished manuscripts, diaries, etc.) of the subjects. Scholarly work of this type usually requires several years to flush out such material that, subject to the erosion of time, yet exists for one person. Obviously such an effort is impossible for the compilation in this book.

Sources include definitive biographies, but as noted above, there are relatively few of them. There indeed are a few autobiographies, though one has to be careful about them. Where possible, however, I have sampled the writing (both professional and nonprofessional) of our botanists, to obtain some feeling for their ways of expression and thinking.

Texts about the history of botany and biology that tell about the work of individual botanists are helpful. But their subjects rarely rise beyond the disembodied spirits mentioned above. (The current text for botany is Morton, A.G. 1981. History of Botanical Science. 474 pp. Academic Press.)

Obituaries are a convenient source of information but have to be taken up with caution. Most tend to exaggerate in glorifying their subjects. "Speak no evil of the dead." And obituaries for the more recent botanists lack the time perspective (often at least 50 years) usually necessary to evaluate their contribution to the mainstream of the discipline. Published reviews of books written by the subject are helpful. They often provide a more qualified light than do obituaries.

References commonly contradict one another in regard to both facts and interpretations. I have had to tread water carefully to avoid perpetuation of legends or misinformation of the past, and probably have not been entirely successful.

A short bibliography (usually 2-4 references) is provided following the text about each person listed.

CHOICES OF BOTANISTS TO INCLUDE

Starting B.C. with Aristotle and Theophrastus, we come to the near present but include no living botanists. As stated above, it is yet too early to evaluate their place in posterity. The primary criteria have been those of major contributions to botany and availability of supporting literature. In many cases nothing biographical has been written about the subjects, except obituaries. I think most of those listed here would be in the Hall of Fame of about anyone familiar with the history of botany. Others, perhaps not so distinguished, are individuals who have contributed materially to plant science in one or several ways, but don't glow like Darwin or, in a different manner, George Washington Carver. And there are more, a sample, not in the competition for botanical knighting, whom I have found different and interesting.

Oh yes, someone may inquire concerning the predominance of males in this book. Response is simple. Until recently there has been but limited, or nil, employment opportunities for women. That has changed, and were this book written ca. 2050, the proportion might be quite different. For example, of the most recent three botanists, two are women.

WESTERN CIVILIZATION?

One reviewer has asked why these botanists "came so overwhelmingly from N.W. Europe and Western civilizations." I suggest that botanical achievements were an integral part of the evolution of the Western civilizations. Morton (1981) has said "The true scientific study of plants—in the sense that we would accept that term today—began as a consequence and as a part of the great intellectual movement which was born in the sixth century B.C. in Asia Minor, in the Greek-speaking cities of Ionia."

THE NONBOTANICAL READER

I have attempted to write this book in a manner that both botanists and others may enjoy the accounts of the labors and peculiarities of this particular brand of people. The problem for those who are not botanists is language. Familiarity with many words (e.g., geotropism, mitosis, parenchyma) really requires the back-

PREFACE   xiii

ground of a botany course that you never had or have now forgotten. Recognizing this, specialized botany terms are avoided as much as possible, and/or are provided in parentheses. I think you (the nonbotanist) will be able to garner maybe 90 percent of that presented. To catch the remainder you might need a dictionary.

🐉 THIS AND THAT

The essays are arranged chronologically. There is an alphabetic index at the back of this book that allows one to find easily a favorite botanist. **Bold face,** used for the first mention of botanists' names within essays, identifies others anointed for individual biography.

One Hundred and One Botanists

Aristotle (384 B.C–322 B.C.)

Aristotle was not a botanist, or at most only indirectly so. Nevertheless his name turns up frequently among the botanists of the herbal period and extends to those immediately following (16th-18th centuries). It is herefore incumbent briefly to attempt him. Among the philosophers of the golden days of ancient Greece, he is perhaps the most universal and encyclopedic. It is not possible even were the writer authoritarian, which I am not, to adequately present him in a short tractate. What I give here is limited primarily to issues that relate directly or indirectly to subsequent botany.

Aristotle was born in Macedon (northeastern Greece), where his father was court physician to the king. His parents died early and he was brought up by friends. Possibly trained for medicine, he went to Athens and was affiliated with Plato's Academy for about twenty years. After Plato's death, he traveled in Asia Minor and Lesbos (an island in the Aegean Sea). Aristotle allegedly married during this period. Subsequently he became tutor of Alexander, who became the conquering Alexander the Great. After return to Athens, ca. 335 B.C., Aristotle established a Lyceum, an institution devoted to study and speculation about any and everything that could come to the human mind. Probably most of his extant writing was done during this time. But after the death of Alexander,

Aristotle found himself in the dangerous political waters. He retired to Macedon and died shortly thereafter.

Reports of Aristotle's appearance and personality, as filtered through the time, are necessarily vague. It is said that he was thin and had a ready wit and tongue. He spoke with a lisp.

I stated above that Aristotle wrote on everything—to specify slightly, ethics, logic and its methodology, physics, cosmology, metaphysics, rhetoric, dialectics, politics, the soul. And biology. And the earth is round.

Biologically Aristotle was a zoologist (or if he also wrote of botany, as some suggest, the manuscripts have been lost). More than likely, however, he left the botany to his pupil **Theophrastus.** Each was emphatically his own man, and the nature of their writing was quite different. Theophrastus is broadly conceptual, direct, and easy to follow. Aristotle is less easily unraveled. At an essential minimum, he was a broad-scope natural philosopher with a bent for logic. Though he was an acute observer, as everybody else, he was not an experimentalist. On the other hand he differed from the others in going beyond the heights of pure reason. He observed first and deduced afterwards. His biology was especially of marine life, descriptive and intuitive as to informal classification; for example, mammals of the land and fishes of the waters were of different groups, but he was smart enough to place dolphins with the mammals rather than fish. It was a long time before anyone agreed that he was correct.

It appears that Aristotle assembled his biological groups by accretion and then hunted for an essence common to all. He was not necessarily consistent in his alleged views and applications of them. For example, when Aristotle was relatively young, he proposed a *logical division* of all knowledge, namely that it should be segmented into successively subordinated paired categories, such as, it is red vs. it is not red. For botanists, this logical division seems analogous to a dichotomous botanical key for plant identification. But this mechanical procedure is illogical for any purpose except identification or convenient categorization. This is because it is stepwise for single characters, whereas affinities must be based on association of multiple characters. Later in life when Aristotle came to his informal classification of animals, he realized this. He depart-

Aristotle

ed from his prior testimony, in fact he "specifically ridiculed dichotomous division as a classifying principle" (Mayr 1982, pp. 150-151).

Biologically, or for the classification of anything else, Aristotle defined causes as material causes, formal causes, efficient causes, and final causes, the latter being completion of object or adaptation. For example, the seed of a maple tree is the material cause and the mature tree, the final cause. I suppose one can quibble with the efficacy of this procedural taxonomy. Does it portray teleology as sometimes asserted?

These examples perhaps provide sufficient reason for varied interpretations of Aristotle in later botanical literature. But he was certainly the most important biological philosopher of his time. Indeed for more than a thousand years following his death, the route was only downwards. Most immediately came Pliny (A.D. 23-79), an uncritical compiler of a disarranged attic of some sense and much nonsense. In contrast, Aristotle's work was based entirely on his own observations and interpretative rationale.

Except for one book (*Organon*, on logic), Aristotle's production was lost to western Europe. His writing, however, was preserved by the Arabs, and began to filter back into Europe at the time of the reawakening, approximately the twelfth century.

And here he filters to you, wherein the selective botanical flash reflection is now faint and blurred. But don't forget that his pupil and successor, Theophrastus, is ofttimes proclaimed as the Father of Botany.

SELECTED BIBLIOGRAPHY

Mckeon, R., ed. 1941. The Basic Works of Aristotle. Random House, New York. 1485 pp.
    Preface and introduction, by the editor, pp. vii–xxxiv. The remainder of the book would perhaps take years to absorb.
McNeil, W.H. 1979. World History, 3d ed. Oxford University Press, New York. 558 pp.
    Primarily pp. 133-148 on Greek civilizations.
Mayr, E. 1982. The Growth of Biological Thought. Belnap Press, Cambridge. 974 pp.
    Aristotle appears on pp. 149-154.
Morton, A.G. 1981. History of Botanical Science. Academic Press, London. 474 pp.
    Aristotle appears on pp. 27-29.
Sarton, G. 1970. A History of Science. Vol. 1. Norton, New York. 646 pp.
    Primarily pp. 522-583.

Theophrastus (ca. 370 B.C.–285 B.C.)

Theophrastus of Eresus has perhaps reasonably been called the Father of Botany because there were none such before him and, for about 1,500 years, none after him. But he cannot correctly be called the Father of Botany, and indeed was almost peripherally a botanist. As we will see.

We know but little of the human existence of the ancients. Theophrastus is said to have been born on the Island of Lesbos (adjacent to the Turkish peninsula in the Aegean Sea). As a young man he traveled widely—a tritism; young men travel, old men become cautious and stay home, except for golf and cruises. Then, Theophrastus, in the golden days of Athens, joined **Aristotle's** Lyceum. When Aristotle departed the country for reasons of political safety (ca. 323 B.C.), Theophrastus became the director. The Lyceum evidently was an educational center, perhaps equivalent to a university, attended by people of presumed expertise and philosophical knowledge, who were directed to leadership. The Lyceum had teachers, a library, a botanical garden, and a museum. Theophrastus there remained the rest of his long life, not only as director but as renowned teacher who had innumerable students, and as a prolific writer. A presumed portrait of him, just another gink with a short beard, which is included in several current books, was made in the 1700s, evidently from some preexisting sculpture.

Theophrastus

Was the sculpture prepared from Theophrastus in life or did the creator later dream up what he might have looked like?

We can perhaps somewhat know Theophrastus if we consider his popularity as a teacher (a magnetic and compelling personality?) and perhaps from the rationale of his writing. There he emerges as a careful, analytical thinker who examines all sides of a question, but often takes no vigorous position in any direction. More information is needed he says. Probably his writing about people is atypical, but it is entertaining. Among Theophrastus' diverse subjects was a series called *Characters,* thirty of them, perhaps stereotypes of those then inhabiting Athens, perhaps Theophrastus' classification of the human race as he knew it. Sarton (1952, p. 549) has reproduced a translation of one called *Superstitiousness,* the first paragraph which follows:

> Superstitiousness, I need hardly say, would seem to be a sort of cowardice with respect to the divine; and your Superstitious man such as will not sally forth for the day till he have washed his hands and sprinkled himself at the Nine Springs, and put a bit of bayleaf from a temple in his mouth. And if a cat cross his path he will not proceed on his way till someone else be gone by, or he have cast three stones across the street. Should he spay a snake in his house, if it be one of the red sort he will call upon Sabazios, if of the sacred, build a shrine then and there.

But in addition to botany and that of suggesting the universality of certain human traits, Theophrastus wrote on a fair piece of everything, some 200 titles (McDiarmid 1976), science, philosophy, literature, history, politics, and you take it from there. Some specific titles and subjects include metaphysics, teleology, *De Igne* (of fire; examining Aristotle's thesis of the elementary substances of nature), *De Lapidibus* (of stones and like things). And *Physicorum Opiniones,* wherein "Theophrastus was no less influential as a historian and critic of science than as a scientist" (McDiarmid 1976).

Enough. All of this should settle the botanical notion that Theophrastus was just a botanist. But, in part and in deed, he was a botanist, and the standard statement about him (among botanists) is that he was the Father of Botany. This is not so. True, he was the only one who wrote of botany as a science, not only in his time but for many centuries. But his botanical writings were largely lost for about fifteen hundred years and only rediscovered when others finally, independently, were beginning to catch up with him. Thus rather than a father, he was a prophet, and a lonely bachelor, because he left neither viable seed nor intellectual children.

What we have of Theophrastus, are two books commonly translated as *Inquiry into Plants* and *Causes of Plants*. The other plant writers of the time were the rhizomatists, plant doctors who gathered plants and told how they would cure what ailed you—their interest in plant science went no further. On the other hand, Theophrastus said, "We must consider the distinctive features and general nature of plants, their morphology, behavior in life, morpogenesis and mode of reproduction" (modern terminology used). And he did. The *Inquiry* deals approximately with morphology and classification, and the *Causes* with physiology and reproduction. McDiarmid (1976) labels the *Causes* as etiology.

Theophrastus' writing, at least in botany, contrasts with that of Aristotle's in being direct and pragmatic, lacking the philosophical underpinnings, as for example, in Aristotle's classification of actual and potential existence portrayed in his formal causes, material causes, efficient causes, and final causes. With Theophrastus, one does not have to struggle with these interpretations of ultimate reality.

Theophrastus' botanical essays are alleged to have been continually revised lecture notes. He described plants, provided terms for their parts, attempting also to relate these to function. Unlike most subsequent writers for centuries, he only slightly attempted to analogize plant parts with those of animals; i.e., he did not write of the plant with its mouth to the ground and assigned it no stomach; instead a plant has roots, stems, leaves, inflorescences, and flowers—those of the grasses are flowers as well as of roses. He recognized in general the difference between monocots and dicots (conventionally assigned to writers in the seventeenth century), and he differentiated some of the major plant groups that we now call families. He described some 500 plants. He certainly was the first ecologist (and perhaps plant geographer) to write of the patterned distribution of plants, of plant communities, and of some of the subtleties of climate, water, and soil that affected their community and geographic locale. His physiology, of course, was speculative, but those who followed him for a millennia and a half could scarcely improve on it until the components of the air were distinguished and the fact that these, as well as water and nutrients from the soil, participate in plant growth.

These are only examples to establish my motif. Theophrastus was a botanist, and by far, the first botanist. He was also much more than a botanist. And sadly, he is not the father of botany, of which the nascent, faltering embryology was initiated independently and much, much later.

SELECTED BIBLIOGRAPHY

Arber, A. 1988. Theophrastus. In: Herbals, 3d ed., p. 357, index entries. Cambridge University Press, Cambridge.
McDiarmid, J.B. 1976. Theophrastus. In: Dictionary of Scientific Biography. Vol. 12, pp. 328-334.
Morton, A.G. 1981. Theophrastus. In: History of Botanical Science, pp. 29-43, 50-51. Academic Press, London.
Sarton, G. 1952. A History of Science. Vol. 1, p. 549. Norton, N.Y.

Dioscorides (Pedanius Dioscorides of Anajarbus) (ca. 20–70 A.D.)

Dioscorides flourished (an expression commonly used by historians, but many of these people did not really flourish) about three centuries after **Theophrastus.** If we may call his *Materia Medica* a botany text, it was the botany for a thousand years, far exceeding any other hall-of-fame claimant. We must qualify botany only to the extent that Dioscorides included a few animal products and minerals purported to have medical effects. The half-life and influence of the *Materia Medica* vastly exceed the botany of his predecessor, the scholarly Theophrastus. And why? It is likely because it, the classic pharmacopeia, spoke directly to the immediate needs and suffering of human life. This stimulated much secondary copying and distribution. Theophrastus' manuscripts, on the other hand, scarcely responded to anyone's living exigencies. In the twilight of an intellectual age, they languished in forgotten corners of the Lyceum library and were lost, at least to the western world, for about 1,500 years.

Dioscorides was a Greek physician with the Roman armies, and that is about all we know about him other than speculation (Riddle 1971). The central core of medicine was knowledge of herbs; plants from which extracts could be obtained or preparations otherwise concocted that could be used to cure what ailed

Dioscorides

you. The skilled physician necessarily knew his plants, where they grew, and when to collect them to prepare his healing solutes, powders, or what have you.

Dioscorides' treatise, descriptions of approximately 500 medicinal plants (plus some for flavorings and ointments) with listing of ailments for which they were said to have curative qualities, was a book we presently call an herbal. It was not the only ancient herbal—a few others were contemporaneous, but Dioscorides evidently felt he could do a better job—"I was not moved to this undertaking by any vain or senseless impulse." Either his 'book' was indeed superior or the circumstances of its preparation and preservation otherwise favored it because, while the competition disappeared from sight, the fame of his multiplied. It was initially preserved and copied (often with variation) by the Arabs. It was illustrated by an unknown artist ca. 500 A.D. and subsequently copied and recopied in Latin. It was indeed *the* herbal until the Middle Ages when the works of the Persian physician, Avicenna (980-1037 A.D.) reached the western world. Then there came a vast outpouring of derivative books that, with the invention of printing, could quickly be disseminated and used. But the continuing influence of Dioscorides is suggested in that it was translated into English by a John Goodyear in 1655. This manuscript, however, sadly lost in a library, passed essentially unnoticed until present century resurrection by R.T. Gunther, who, retaining the English of the period, edited copy that was published in 1933. (Riddle 1971, p. 122, says this translation was "woefully inadequate." But it's all I and most of you can read.)

Well, what do we presently see? I include sample pages with text about four plants and an illustration of one of them to give an idea. I am not competent to judge the alleged curative merits of the individual plants and therefore pass them by. The identity is something else. Undoubtedly Dioscorides knew his plants, but no vocabulary existed for presenting identifying characters, and he made no attempt to develop such. Thus his species are portrayed in a mess of verbiage that seemingly would have been but minimal help to one who was not already reasonably familiar with them. Some kinds can be recognized quickly by their illustrations, but in other cases, the figures and descriptions obviously are not of the same thing.

Numerous individuals from the Middle Ages on have struggled with identity of the recondite kinds, and their names, as represented by the binomials given in Gunther's book, are mostly educated guesses.

Sample page from Dioscorides' *Materia Medica*.

I suggest that much of the lore about the plants concerned was passed from generation to generation verbally and by practice, and that the *Materia Medica* primarily served as the codification bible of what most good physicians already knew.

Beyond the descriptions and listing of medical properties, the only explicit quasi-botany is in Dioscorides' short introduction, for example, "Before all else it is proper to use care in sorting up and in the gathering of herbs. We ought to gather herbs, each in its due season, when the weather is clear. The place also makes a difference; whether the localities be mountainous and high, whether they lie open to the wind. It must also not be forgotten that herbs frequently ripen earlier or later according to the characteristic of the country and the temperature of the years." Also Dioscorides did

Dioscorides

some, perhaps unconscious, botanical classification; i.e., similar kinds, as for example, the mints and umbellifers were mostly placed together in textual sequence.

But it is so easy to be critical of those working without Microsoft Word, photoduplication, and National Science Foundation grants to facilitate production. The *Materia Medica* was the direct ancestor of the medieval herbals, which in turn were the progenitors of botanical floras. These plus the early Renaissance fermentation of experimental and mathematical science constituted the genesis of an independent natural science, botany.

SELECTED BIBLIOGRAPHY

Arber, A. 1988. Dioscorides. In: Herbals. 3d ed. (reprint of 2d ed., 1938, with biographical annotations by William Stearn), pp. 8-12 and elsewhere. Cambridge University Press, Cambridge.
Gunther, R.T., ed. 1933. The Greek Herbal of Dioscorides. Oxford University Press, Oxford. 701 pp.
Illustrated A.D. 512; English translation by J. Goodyear A.D. 1655.
Morten, A.G. 1981. Dioscorides. In: History of Botanical Science, pp. 66-68, 86-88. Academic Press, London.
Riddle, J.M. 1971. Dioscorides. In: Dictionary of Scientific Biography. Vol. 4:119-123.

Hildegard of Bingen (1098–1179)

The writing about plants between 1100 and 1650 was mostly that of medicine and plants. The manuscripts and books produced during this period are collectively called the herbals. The first herbalists (those who wrote the herbals) were primarily copiers and translators of **Dioscorides** and **Theophrastus**. The early herbals were handwritten and illuminated (meaning with lots of fancy artwork, symbolic do-dads, mystic background, and the like). They told what plant cured what ailment, and some related the uses of plants for food and otherwise. The object of descriptions, of illustrations and, in some instances, rudimentary classification, was hopefully to enable one to identify the plants. The invention of printing (Gutenberg, ca. 1454) opened a whole new world in production; illustration evolved from crude wood engravings to high-quality copper plates. Written texts changed from dependency on ancient manuscripts to the inclusion of new knowledge based on the plants themselves.

The many herbalists included no one or two dominating figures. Among the horde, perhaps you have heard of some if the better known names: **Gesner, Gerard, Bauhin, Cesalpino,** Fuchs, **Camerarius, Mattioli.** But you probably have not heard of Hildegard. Neither had I until reading Anderson (1977).

Although botany was just a tiny piece of her vast output,

Hildegard of Bingen

Hildegard was possibly the first woman of record to write about plants. She was the offspring of nobility; Anderson says her father was a knight. She was born in what is now Germany, and placed in a convent at age eight where she became abbess at thirty-seven. Twelve years later, her reputation having grown beyond her facilities, she established another convent where she remained as bosswoman, the abbess, until her death. She was a gifted mystic, i.e., was always seeing visions that indeed, I conjecture, included directions from the deity. She corresponded with emperors, kings, and popes to the effect that the church was the supreme administrative unit and all owed first fealty to it, i.e., "she was a staunch papist . . . opposed imperial encroachments, defended clerical privileges" (Newman 1987). Nearly all of her manuscript works were of theological nature, mysticism, and interpretation of religious dogma. Beyond her influence on the religious thought of day, she was a feminist who was death on sex, which she regarded as derivative of Original Sin. It has been suggested that her writing was influential on the early evolution of the German language (if there are yet graduate students whose Program of Study specifies German, they may feel this is not to her credit).

Hildegard was plagued by ill health and writes of periods of bed confinement, weeks at a time. But like **Darwin,** also a semiinvalid most of his life, this did not stop her management responsibilities or her persuasive, often polemical writing. In fact perhaps her illnesses abetted her reputation, for it now seems agreed that her visions were probably pathological or physiological deriving from unknown infirmities.

But this is a long way from Hildegard as an herbalist. Indeed in contrast to Hildegard's status as one who talked with God, the *Physica* (Natural History) and *Causae et Cures* (Causes and Cures) concentrating on the here and now of economic botany, i.e., folk medicine, food, and everyday living, might be viewed as a wild card. But these represent the justification for considering her a protoherbalist. Their origin and Hildegard's interest possibly derive from the frequent role of the abbeys and convents as an extension service, especially for medical help, for the adjacent population. The *Physica* includes some 200 short essays on plants and medical use. The rationale of the *Causae* is based on the four humors, derived I suppose, from Hippocrates, fanciful to us, but primarily of naturalistic basis. What is good for what, how to prepare, dietary advice, gynecological problems. For both books, one might say Hildegard collected the cures of the past, the remedies of the present, and blended them with her own observations and intuition. Concerning food she said that beans were better for humans than peas (but what kind of "beans?" The Old World did not yet have the American *Phaseolus*). Medically she wrote of what was available, for example hempseed (only the seed? hemp is marijuana, you know), rose leaves, nutmeg, Hymelsluszel (*Primula*), Storcksnabel (*Geranium*), and Hufflatich (*Tussilago*). She even provided an explicit description of baking bread of which Bartholomew said a century later "and at last after many travails, man's life is fed and sustained therewith." (This enumeration is from Anderson 1977.)

Previously I said Hildegard collected the cures of the past. These were ones known to her from the folklore of the times. There seemingly is no sign that she consulted Dioscorides or any other past writers. As with other topics, everything came directly from her experiences, acquired knowledge, and views about it.

The original Hildegard *Physica* manuscript, probably written about 1150, is not known. The earliest printed edition is from 1533. It is slightly illustrated by woodcuts that, however, had nothing to do with the text itself (Anderson 1977).

Hildegard at a minimum was a brilliant and assertive person. She was proposed for canonization. But, evidently the case was ill-prepared and slovenly; it languished in the procedural bureaucracy without affirmative action. One wonders, had she been born of the present, how she would have reacted to the present liberalization strife of the Catholic Church and what she would have done about it.

SELECTED BIBLIOGRAPHY

Anderson, F.J. 1977. The *Physica* of Hildegarde of Bingen. An Illustrated History of the Herbals, pp. 51-58. Columbia University Press, New York.
Flanagan, S. 1989. Hildegard of Bingen, 1098-1179. Routledge, London. 230 pp.
Newman, B. 1987. Sister of Wisdom. University of California Press, Berkeley. 189 pp.
Singer, C. 1928. The Visions of Hildegard of Bingen. In: From Magic to Science, pp. 199-239. Ernest Benn Ltd., London.

Otto Brunfels (ca. 1489–1534)

Brunfels was the earliest of the three herbal-producing "German fathers of Botany" of **Sachs'** *History*. If by fathers, Sachs meant German priority, he was claiming too much ground; the Italians really were first. Anderson (1977) has said, "The modern age of botany began in 1530 when Brunfels issued his *Herbarum vivae eicones*." This is too much. Brunfel's text was often less than appealing, and his concepts were befogged with misunderstanding. Much of the credit for the success of the *Herbarum* belongs to the contributing artist, Hans Weiditz. But Brunfels was a nice guy, and industrious as hell—if that is any help.

Born in Switzerland, Brunfels completed his formal education with an M.A. at Mainz. A subsequent stint in a monastery terminated as a consequence of his becoming an ardent follower of Martin Luther. Apparently he was then an itinerant Protestant minister for a few years. Then he settled in Strassburg, supporting himself as a schoolteacher, writing, compiling, editing, and translating. His subjects, though various, were primarily of medicine, medical plants, and pharmacology. But he also wrote theological treatises, and even one on the evils of astrology. He subsequently moved to Basel, there receiving an M.D. (1533) almost at the end of his life. His terminal residence was in Bern as town physician.

Brunfels married in 1524. His wife, Dorothe Heilgenhensia,

saw his latter writing through publication after his death.

And about the quality and impact of the *Herbarum Vivae Eicones* again. Joining Anderson, Stannard (1970) says that it was "a book destined to change the direction of botany." In terms of the illustrations, maybe so, but I still think this is going it a bit strong. **Arber** (1986) more carefully states that his work "initiated a new era in the history of the herbals," which perhaps is reasonable.

In a positive vein, Brunfels introduced two innovations soon followed by other herbalists. First, the illustrations (woodcuts) were of a quality never before achieved. Indeed these, of the artist-engraver Weiditz, were subsequently praised to the extent that one feels that Weiditz rather than Brunfels should be the subject of this essay. Weiditz worked with "plants from nature" rather than copying prior, often ancient, stylized, and tired woodcuts dating back to those prepared for **Dioscorides'** plants. All in all, it was a case of the "pictorial tail wagging the textual dog." (Anderson 1977).

The second innovation in the *Herbarum Vivae* was that Brunfels did not try to identify all of his plants with those of Dioscorides (then almost an act of heresy), but listed some German kinds that could cure what ailed you. **Theophrastus,** more than fifteen hundred years prior, had said that plants of one region differ from those of another. But early herbalists read Dioscorides, not Theophrastus, and seemingly assumed that those of the medical god were universal. Brunfels by indirection said this was not true, in that he listed some German medical plants not included by Dioscorides. And he also described some German plants independent of their medical merits. Here was the germ of the local or regional flora idea—the divorcing of botany from medicine.

Such is a synopsis of the botanical portion of Brunfel's curriculum vita. Certainly he can be credited for obtaining the collaboration of Weiditz, and then leaving him alone to do his thing. Negatively, his descriptions, supposedly the soul of the book, were too often pilfered or adapted from inadequate earlier writing, not from plants. They were evidently often second rate.

Where does this leave us? Let us say the *Herbarum Vivae* bearing Brunfel's name constituted a spirit for the future. This is supported by an account (Arber 1986) that he walked forty miles in his latter years to get acquainted with the younger and modest **Boch**

Otto Brunfels

(another of the German three), and encouraged him to write his herbal from his garden and collections. Brunfels is remembered in *Brunfelsia* (Solanaceae).

SELECTED BIBLIOGRAPHY

Anderson, F.J. 1977. The Herbarum Vivae of Otto Brunfels. In: An Illustrated History of the Herbals, pp. 121-129. Columbia University Press, N.Y.

Arber, A. 1986. Brunfels, Otto. In: Herbals, Their Origin and Evolution, 2d ed., reissue with annotations. Cambridge University Press, Cambridge. Numerous entries in index, p. 341.

Stannard, J. 1970. Brunfels, Otto. In: Dictionary of Scientific Biography. Vol. 2, pp. 535-538.

Luco Ghini (1490–1556)

In science, nearly all whose names remain reflected in the corridors of posterity are those who express new ideas in publication. Ghini published nothing of botanical substance. Yet because of his role as the supreme teacher and promulgator of botany in his time, and his practical innovations (the herbarium, the botanical garden), he belongs among the anointed.

Ghini, Italian, studied medicine at Bologna. He was soon (1527) lecturing on "simples" (plants used medically) and in due time advanced to professorship. He moved to Pisa in 1544, returning to Bologna the last two years of his life.

I read that Ghini's father was a notary (one who attested to the rectitude of legal documents). Ghini married in 1528, and there is incidental mention of a son. That's all, for nonbotanical life.

After going to Pisa, Ghini maintained his home in beloved Bologna where he had his own garden in which to play, there often accompanied by some of his students. From the nature of his accomplishments, I have imagined him as an unselfish, enthusiastically benevolent, and magnetic person. Prior writers speak of his "wide culture . . . selfless nobility of character . . . attractive personality . . . enthusiasm and scholarship."

Plainly, Ghini was a revered botany teacher. Most of the bud-

Luco Ghini

ding herbalists, naturalists, and botanists of the time were at least briefly his pupils. Among these, **Cesalpino** became the most important in the next generation. Morton (1981) says that he directed botany away from scholasticism to plants and that "the rise of botany in the first half of the sixteenth century to become a field of genuine study owes more to Luco Ghini than any other individual."

Ghini helped others. The outstanding example is **Mattioli** who was hell-bent on trying clearly to identify **Dioscorides'** plants. He asked Ghini to participate. Ghini became excited about it and traveled about the Mediterranean countries and Near East (Dioscorides' area) looking for species that would match the classic but unsatisfactory descriptions. Possibly he intended to publish his findings, or some of them, but all were absorbed in Mattioli's *Commentari* on Dioscorides. The only Ghini carryover from this was a thing called the *Placiti* (published posthumously) that, I gather, was a travelogue of his expeditions.

Ghini was the first of record to make a herbarium (*hortus siccus*) or if not, he was the first to recognize its usefulness and to put the idea into practice. He assembled his own collection; he sent specimens to Mattioli; and he probably demonstrated the technique in his teaching. The procedure was simply that of drying plants under pressure between sheets of paper and then gluing or otherwise fastening them to pieces of cardboard. These could then serve for permanent reference and be exchanged with other botanists. In retrospect, this innovation seems almost absurdly simple, but that is the way of many novel inspirations. The word herbarium in its present sense comes from **Tournefort**.

Except for medical herbs, the convenience of having live plants growing close to home and easily available for study was still lacking. Ghini established the botanical garden at Pisa; one of his students listed some 600 species in it. Maybe it was the first, or maybe that at Padua was prior. It really doesn't make any difference. Again, an innovation caught on and soon there were botanical gardens all over Europe.

Among several references, I have not seen a single uncomplimentary statement about Ghini. That is unique. I wish I could chat with him.

SELECTED BIBLIOGRAPHY

Arber, A. 1988. Ghini. In: Herbals, 3d ed., p. 347, index entries. Cambridge University Press, Cambridge.
Keller, A.G. 1972. L. Ghini. In: Dictionary of Scientific Biography. Vol. 5, pp. 383-384, with citations; all in Italian.
Morton, A.G. 1981. Ghini. In: History of Botanical Science, pp. 120-123, 153. Academic Press, London.

Hieronymus Boch (1498–1554)

Inconsistencies in references about Boch start with his name. Apparently he was born as Jerome Boch, but wrote as Hieronymus Boch (or Bock) or Hieronymus Tragus (Latin translation of Boch). He followed **Brunfels** among the trio of German herbal fathers of botany. He was the opposite of Brunfels in the character of his *Kreuterbuch* (all plant authors of this time had their *Kreuterbuch* or its equivalent). The first edition was unillustrated (he learned his lesson; later editions included illustrations), while those of Brunfels prepared by Weiditz were supreme. On the other hand, Boch was for real; he was a knowledgeable botanist who wrote trenchant descriptions, while Brunfels' efforts were often sad adaptations from prior writers.

The information about Boch's life is scanty and ambiguous. Evidently there is no record of his education. Since most of his years apparently were spent as a Lutheran pastor, this presupposes theological training, and hence, probably university experience even if he finished no degree. There is no record of such.

Boch's botanical turn probably came from nine years as caretaker of the gardens of a Count Palatine Ludwig. His life thereafter was double—that of a minister and simultaneously a zealous explorer and compulsive writer of botanical wonders. No, not double, triple—in 1523 Boch married Eva Victor. Also, in a day of religious zealotry, he had to do some moving around because he was an apostle of Martin Luther. That is about the sum total of personal information.

In preparing the *Kreuterbuch* (first edition, 1539), Boch said his objectives were to treat German plants, provide their names, their characters, and medical virtues. He pointedly abandoned **Dioscorides** and prepared his own sequence for the text, which was, in effect, a convenience classification of some 700 plants. Crude yes, but a classification, an attempt to consider similar kinds together. All, except for Morton (1981) who have written of him, presumably from firsthand knowledge of his productions, say the descriptions were supreme. For example, he was one of the earliest to observe and characterize stamens and pistil(s) of flowers (yet minus function, of course). Morton, however, states that the descriptions were often naive and far from systematic. So we need to pick up a copy and decide for ourselves—no easy task; the incunabula of this era are now few and far between. Of the people here cited, I know only that Frank Anderson (1977) had access to one or several editions of the originals and could face up to the German and Latin in them. Be that as it may, I pass with the suggestion that maybe Morton shrugs Boch off too casually.

Boch, who traveled around the German countries looking at plants, may have been the first real field botanist following **Theophrastus** except for Albertus Magnus, the "bishop with boots," a couple of centuries earlier. The results of Boch's field knowledge are evident in ecological, distributional, and phenological notes along with the descriptions. Examining the ancient and persistent superstitions that larded folk lore about plants, he discarded or questioned paranormal phenomena.

Despite these merits the *Kreuterbuch* drew but little popular attention as compared to the Brunfels/Weiditz. It became a good, if not best, seller only with subsequent editions for which he obtained funding to employ an artist who prepared woodcuts.

Perhaps exaggeration, but I am willing to settle for Anderson's (1977) assay of Boch: "An amateur . . . largely self taught . . . his accomplishments moved the study of plants closer to a science . . . than any . . . since Theophrastus . . . 4th century B.C." The genera

Hieronymus Boch

Tragus (Gramineae) and *Tragia* (Euphorbiaceae and several others in this same family) are named for him.

SELECTED BIBLIOGRAPHY

Anderson, F.J. 1977. The Kreüter Buch of Hieronymous Boch. In: An Illustrated History of the Herbals, pp. 130-136. Columbia University Press, New York.
Arber, A. 1986. Boch. In: Herbals, Their Origin and Evolution, 3d ed., pp. 55, 58-59, 61, 151-152, 166. Cambridge University Press, Cambridge.
Morton, A.G. 1981. Boch. In: History of Botanical Science, p. 125. Academic Press, London.
Stannard, J. 1977. Boch, Jerome. In: Dictionary of Scientific Biography. Vol. 2, pp. 218-220, with citations.

Piendra (or Pier Andrea) Gregorio Mattioli (1501-1577)

To sample the 500-year span of the herbals, we now proceed to Mattioli. Perhaps I am especially interested in him because I have the good fortune to own a Mattioli heirloom. This was a period that included the multiple genius, Leonardo da Vinci, Copernicus (the earth goes around the sun), Vesalius (human anatomy), Francis Bacon (philosopher of science), and William Harvey (circulation of the blood). Botany, then describing and illustrating plants and telling what they were good for, lagged far behind. Not until we come to **Van Helmont** (1577-1635), an alchemist who grew a willow tree (that gained weight but the soil did not lose weight), do we broaden the vision of plants.

Mattioli was Italian. His classical education was followed by an M.D. at Padua in 1523. He practiced medicine in several places and associations, rapidly coming to high repute. For he spent some years in the court of the Emperor of the Holy Roman Empire, caring for the welfare of that worthy and his successor and their families. And along the way he was married thrice, producing several children. But he is scarcely remembered by this fact, nor even that he tended the ailments of an emperor. Rather, the emperors evidently were sufficiently healthy that the doctor had the time to do what otherwise seized him. Mattioli published several medical books, a long poem, and a translation of Ptolemy's *Geography*. But his fixation was **Dioscorides;** he was determined to be the world authority on Dioscorides and to bring him up-to-date in light of the new knowledge of a millenia and a half. His production, *Commentarii . . . Dioscorides* was a bestseller. It went through several editions, of which the initial ones are said to have sold 32,000 copies (for then, fantastic), and its unprecedented distribution made him famous all over Europe. The title *Commentarii* was misleading. True Mattioli took up and identified all of Dioscorides' plants (and tolerated no question of his accuracy) and expanded his descriptions. But the descriptive text was yet crude—Mattioli pioneered neither in the development of an adequate terminology nor in classification analysis. However, he innovated in two ways. First he listed a few new plants that had nothing to do with Dioscorides nor indeed with medicine. This was a precedent that stimulated the evolution of the herbal from a medical-cure book towards a floristic listing descriptive of the European plant world—botany beyond the umbilical cord of its curative powers for warts or other ailments to which humans are subject. The floristic concept indeed gradually unfolded in subsequent herbals culminating in **Bauhin's** *Pinax* that included ca. 6000 species. Secondly, Mattioli's treatment was illustrated with hitherto unexcelled woodcuts that allowed recognition of the plant even though the descriptive text was murky.

Initially Mattioli may have been a pleasant fellow, and most historians have left him in that light. But after becoming famous, he evidently developed cancer of the ego, and tolerated no criticism or disagreement from others. He lambasted anyone who had the temerity to disagree with him, and his reputation was sufficiently formidable that he ruined several who took issue with him. The titles of his *Apologia Adversus* and the *Censura* are perhaps sufficient to suggest the nature of his bigoted wrath. It is the recent research of Frank Anderson (1977) that has reasonably documented Mattioli's personality characters, and Anderson, usually so kindly, even evaluates Mattioli's competence in less than a complimentary vein.

One of the waves of the plague finally terminated Mattioli, for which Anderson seems to breathe a sigh of relief. But Matiolli's *Commentarii* went through fourteen editions and was subsequently translated into Latin, German, French, and Czech. Who can quarrel with success and fame?

SELECTED BIBLIOGRAPHY

Anderson, F.J. 1977. The Commentarii of Pier Andrea Mattioli. In: An Illustrated History of the Herbals, pp. 163-172. Columbia University Press, N.Y.
Arber, A. 1986. P. Mattioli. In: Herbals, Their Origin and Evolution, p. 351, index references. Cambridge University Press, Cambridge.
Zanobio, B.P. Mattioli. In: Dictionary of Scientific Biography. Vol. 9, pp. 178-180, with citations.

Valerius Cordus (1515-1544)

Cordus belongs among the immortals even though his short life only allowed him to get started. He was the first to look at flower parts and inquire, "What do you have to say?"

Cordus' father is said to have been a botanist, physician, and poet. The son, after obtaining a B.A. at Marburg, went to the University of Wittenberg. Already an expert on the mysteries of **Dioscorides,** he gave lectures on that classic doctrinaire of plant medicine. At Wittenberg he also became familiar with the scattered and diverse pharmaceutical literature of the time. He traveled extensively and wrote plant descriptions. Cordus died at the age of twenty-nine following an accident and "fever."

Nevertheless, like Schubert's unfinished symphony (Schubert died at thirty-one), Cordus left beauty behind. Apparently he was both a genius and a nice fellow—an uncommon combination. He possessed a "singular brilliance of intellect and charm of character"; he had an "appealing personality"; "he was spectacularly gifted." Clearly quite a chap.

Cordus' untimely death precluded publication, but he had been writing madly his entire adult life. He left behind stacks of manuscripts which (great good fortune!) were mostly rescued by the Swiss botanist **Konrad Gesner**. And Gesner promptly served as editor and publication expediter. I have several titles in front of me but I won't bother to list them beyond abbreviations of the polynomial titles of two, the *Dispensatorium* (Pharmaceutical) and the multi-volume *Historiae* (of plants). What I want to talk about is his botanical writing.

Among the incunabula herbals, the descriptions of the plants listed were usually the weakest parts of the text. It is impossible to identify many, or sometimes most, of the plants from the casual commentary about them. Surely the authors, who presumably knew at least some of them firsthand, could have done better. **Agnes Arber** (1986, p. 146) has suggested the provocative hypothesis that knowledge of the identity of plants was mostly passed from generation to generation by word of mouth and demonstration. The descriptions were intended for no more than helpful labels. This supposition implies that the written "descriptions" were, in effect, names for plants that the herbal doctors already knew. Perhaps one could identify an individual as "fat man with cigar." While this is scarcely a complete description, were it known that he was the only one of this kind it would serve to identify him. Then one could go ahead relating his virtues or sins, the important part.

But if this was so, Cordus would have none of it. He was emphatic in the belief that descriptions must be diagnostic, and indeed his descriptions drew unanimous applause. He seems to have been the first to look at flowers in detail and to espouse the idea that details of the flowers were and are more important in establishing the relationships and identity of plants than the more conspicuous vegetative characteristics. Hampered by the paucity of words to express what he saw, he developed an elementary phytography. For example, he distinguished among the four whorls making up a complete flower as *calix* (calyx), *foliolus* (petal), *stamina* (stamen), and *apex* (style).

In comparison with his contemporaries, Cordus as a botanist went well beyond the needs of a plant-medical handbook. And all of his botany was from his own observations and knowledge; it was not blindly compiled from earlier sources. One author has said, "There was **Theophrastus;** there was nothing for 1,800 years; then there was Cordus." He is slightly commemorated in *Cordia* L. (Boraginaceae).

SELECTED BIBLIOGRAPHY

Arber, A. 1986. V. Cordus, In: Herbals, p. 343, index entries. Cambridge University Press, Cambridge.
Greene, E.L. 1909. V. Cordus. In: Landmarks of Botanical History. Smithsonian Misc. Publ. 54, pp. 270-314. Smithsonian Institute, Washington, D.C.
Harvey-Gibson, R.J. 1919. V. Cordus. In: Outlines of the History of Botany, pp. 14-18. Black, Ltd., London.
Schmitz, R. 1971. V. Cordus. In: Dictionary Scientific Biography. Vol. 3, pp. 413-415, with citations, mostly German.
Sprague, T.A. and M.S. Sprague. 1937. The Herbal of Valerius Cordus. J. Linn. Soc. Botany (London) 52:1-113.

Rembert Dodoens; Dodonaeus
(1516–1585)

The European Low or Flemish Countries constitute the present Belgium and the Netherlands. They contributed three major figures to the herbals, Dodoens, **Charles de l'Eluse (Clusius),** and de l'Obel. Because they were in constant contact, it is hard to know who to credit with what. Dodoens was the senior of the group.

Born in Mechelen, then of the Netherlands, now Malines, Belgium, he was originally Rembert van Joenckema. He received his M.D.-equivalent in 1535. He was then peripatetic for about eleven years, becoming an internationally known physician and writing on a variety of subjects. Then he came back to Mechelen and was town physician for some years. He had two wives and five children. During a period when the Flemish were fighting for independence from Spain, the Spanish armies destroyed the town. Dodoens fled, accepting an offer as personal physician to the emperor in Vienna, taking his family with him, one hopes. But he didn't like court life in Vienna and wished to go home. It took him two years to get there, but in 1582, he accepted a position as medical lecturer at Leiden University, on condition that he stay out of current religious controversy. There, though internationally known both in botany as well as medicine, his assignments were entirely

Rembert Dodoens

medical because there was yet no botany chair at Leiden. The teaching of botany and development of the famed Leiden botanical garden had to be left to his successors (Clusius and l'Obel).

Ignoring an early epistle on cosmology and astronomy, Dodoens first book was *De Frugum Historia* (vegetables, grains, animal food). Then in 1554 came the *Cruÿdeboek,* evidently the first of its kind in Dutch. With 900 pages, 1,309 woodcuts, descriptions with "color and poetry" (possibly in part borrowed from Fuchs), it became both popular and widely used in his country. True the descriptions and glossary were less than ideally precise, but this was mostly par for the course. The *Cruÿdeboek* was really *the* Dodoens book, because nearly all that followed, terminating in 1588 with the *Stirpium Historiae Pemtades Sex Sive Libri* (colloquially the *Pemtades),* were mostly revisions and improvements of the original.

Dodoens was translated into several languages. An English version by Henry Lyte is said to have supplied the basis for Shakespeare's occasional botanical references. See **Gerard** concerning his (Gerard's) possibly illicit affair with the *Pemtades.*

Medically, Dodoens said that the then popular Doctrine of Signatures was nonsense. Botanically his contributions are hard to define. The definition of a flower as given in the glossary of terms (in the *Pemtades*) is as follows: "The flower we call the joy of trees and plants. It is the hope of fruits to come, for every growing thing, according to its nature, produces offspring and fruit after the flower. But flowers have their own special parts." One can scarcely disagree with this semi-poetic functional definition, but it scarcely supplies confidence in Dodoen's skill in phytography.

However, Dodoens brought together the members of various plant groups now recognized as families such as the Geraniaceae, Compositae, Umbelliferae, and Gramineae. So it is stated. Some authors say that he was the first to do so, but I think this is scarcely true. Seemingly he had no guiding principles in classification, either stated or implicit. For example, the English translation includes six "books," their contents defined in inconsistent ways. Book One, dealing with herbaceous plants, is arranged alphabetically. Book Two is an irregular assemblage based on utilization, i.e., flowers used for garlands, bouquets, and aromas. Book Six has the messy title, "Shrubs, Trees, Forest Trees, and Evergreens."

If there were significant delineations of plant families or generic groups, it must have been within these amorphous headings.

But no matter. Dodoens plowed virgin soil subsequently more finely tilled by his compatriot successors, l'Obel and Clusius. And the effervescent inscription on his tomb in the Church of St. Peter in Leyden begins "To an excellent man of the Greatest Worth . . ."

SELECTED BIBLIOGRAPHY

Anderson, F.J. 1977. Rembert Dodoens. In: An Illustrated History of the Herbals, pp. 173-180. Columbia University Press, New York.

Arber, A. 1986. Dodoens. In: Herbals, 3d ed., p. 345, index references. Cambridge University Press, Cambridge.

Florkin, M. 1971. Dodoens. In: Dictionary of Scientific Biography. Vol. 4, pp. 138-140.

Konrad Gesner (Conrad Gessner) (1516–1565)

Gesner (*Gesnaria* L., Gesnariaceae) was Switzerland's man among the herbalists who were trying to be botanists. He was apparently the first of several generations of a prolific Gesner family who became important figures in the natural sciences, the humanities, and theology.

References seen are somewhat in conflict concerning Gesner's life and botanical productions. Clarification would require research beyond that for this informal series. The story given here is primarily derived from **Arber** (1986) and Morton (1981), both professional historians of the history of botany, with the reservation that the Morton is possibly derivative from the earlier meticulous Arber. The most discursive botanical reference about Gesner is that of Greene (1983), but that author's circuitous route and vision of the forest is obscured by his preoccupation with the trees. Because the formal publication dates of some of the citations given here given may be puzzling, explanatory annotations are provided in the bibliography.

Gesner presumably became enamored with plants as a youngster in his uncle's herb garden. He had a good elementary classical education. Then he became a super compiler, a jack-of-all-learning, a linguist (e.g., he wrote a Greek-Latin dictionary), theologian, physician, zoologist, and peripherally a geologist. And a botanist.

Prior to obtaining an M.D. at Basel (1541), he was a professor of Greek at Lausanne. Then he went to Zurich for the rest of his days, a professor of philosophy before inheriting the chair of natural history.

Reference to the personal life of Gesner is scant. Greene (1983) says that he was married at the age of twenty, and that he then opened a school for boys to support himself and wife, and studied medicine. I have seen nothing about progeny, but presumably there were some since he had descendants (e.g., Johannes Gesner, a contemporary of **Linnaeus**). Gesner is credited with poor health, but this allegation must be qualified by his spectacular writing energy. He is said to have been poverty stricken most of his life. But he both held a professorship, was a practicing physician the latter half of his life, and some of his writing brought in money. Perhaps his poverty was a consequence of a large family?

Now about Gesner's writing and editing. Of the latter, Gesner rescued the voluminous unpublished material of the brilliant **Cordus,** saved it from oblivion and saw it through publication, a major contribution itself. Amid a myriad of other writing and publication, a *Bibliotheca Universalis,* four volumes (1545-1555), said to be an index to all writers on all topics "earned him fame . . . and brought him into correspondence with all of the scholars of his time," (Pilet 1972). Then he did an encyclopedia of the animal world, *Historia Animalium.*

Although Greene (1983) discusses numerous botanical publications of Gesner's, apparently botany in a major way came to Gesner primarily in the latter part of his life. He undertook an *Opera* (or *Historia*) *Botanica,* purportedly to be a survey of the plant kingdom similar to the *Animalium*. In addition to text, he prepared some 1,500 illustrations himself. Then he died of the plague.

Then what happened? Pilet (1972) says the material was published "in two volumes 1551-1571" but provides no citation, beyond crediting them to a C. Schmiedal (note that 1551 was well before Gesner's death). Two others (Arber 1986, pp. 110-113; Morton 1981, pp. 127-128, 156-157) assert that his materials disappeared, and that their whereabouts for some 400 years yet remains an only partly solved mystery. Gesner, they say, willed his materials to a

Konrad Gesner

friend by the name of Wolf, requesting that they be prepared for publication. Wolf failed except for a few illustrations. Next, it seems that **Camerarius** had at least some of the illustrations because he published about forty of them shortly after Gesner's death. Nothing then is known for nearly 200 years until a Dr. Trew published some of them (1751, 1771). But they were so reduced that the details (especially of flowers) were lost. Then oblivion again until 1929 when nearly everything was found in a library attic at the University of Erlangen, the drawings, notes, and all. Morton (1981) says that these are now in the process of facsimile publication.

The papers of most of us could disappear for 400 years and would never be missed. That is not true of Gesner. His historical virtue owes in part to the fact that he was preceded only by Cordus in recognizing the importance of flower, fruit, and seed details in classification. Since he had previously edited Cordus' manuscripts, the idea possibly is derivative. Be this as it may, he prepared a real classification (natural groups within groups) supported by his magnificent illustrations. The scope of his would-be *Opera Botanica* was a significant stage in the graduation from plant medicine towards classification and floristics.

However, if Gesner had seen the Virgin Mary come to earth in the sixteenth century, but this was not announced until the latter nineteenth century when numerous other sightings had been recorded for 300 years, what real historical significance has his vision?

I suppose one responds in several ways. Assuming that Morton and Arber are reasonably correct, there is the matter of credit where it is due. I suspect some of you will say what the hell difference does it make now. However, if next year, manuscripts of a deceased botanist were discovered showing that herm (he or she) had proposed the DNA structure a couple of years before Crick-Watson, I am sure it would be vigorously noised about. And remember also that Gesner was not anonymous in his time, his theories being expressed in voluminous correspondence with others—**Bauhin** published some of them in 1591. Certainly **John Ray,** foremost in systematics in the seventeenth century, saw them and Gesner may have influenced his thinking. Enough was and is known of Gesner's systematic proposals that his name deservedly is maintained in botanical history even if consistent documentation about it is yet lacking.

SELECTED BIBLIOGRAPHY

Arber, A. 1986. K. Gesner. In: Herbals, 3d ed. Cambridge University Press, Cambridge.
   Gesner appears on pp. 110-113, also Index, p. 347; although this book is labeled as 3d ed. of 1986, it is a posthumous reprint of Arber's 1938 2d ed., plus annotations by W. Stearn; hence Arber's writing is of 1938 and prior to Morton.
Greene, E.L. 1983. K. Gesner. In: Landmarks of Botanical History, Part II, pp. 747-797. Stanford University Press, Stanford.
   Greene died in 1915; thus the citation date is misleading. The book is derived from handwritten text among the author's papers, that were edited, and prepared for publication, 1983, by F.N. Egerton.
Morton, A.G. 1981. K. Gesner. In: History of Botanical Science, pp. 127-128, 156-157. Academic Press, London.
Pilet, P.E. 1972. K. Gesner. In: Dictionary of Scientific Biography. Vol. 5, pp. 378-379.
   Citations of the subject's writing, usually given in this dictionary, are lacking.

Andrea Cesalpino (Andreas Caesalpinus)
(1519–1603)

Morton (1981) devotes fifteen pages to Cesalpino, considerably more than to any other botanical luminary except **Ray,** who came about 100 years later. While I don't necessarily regard this as a measure of Cesalpino's preeminence, plainly he was somebody. A simplistic rationale might be that he, as the preeminent botanist of his time, more than anyone else represents the cleavage between the herbalists and the botanists. It is indeed alleged that he was really the first important 100 percent botanist in 1,700 years, i.e., since the Greek **Theophrastus.** He did not write about medical plants, which were the raison d'être of the herbals. Instead, he wrote a botany book giving the principles of botany as seen by him in his time and presented a classification of plants, which was more precisely honed than that of any predecessor. He was "it" for 100 years for anyone who cared to look, and certainly Ray and **Linnaeus** did so. Linnaeus, usually preoccupied with himself, had warm praise for him.

Cesalpino, Italian, studied medicine and botany at Pisa under the famed teacher **Luco Ghini.** After getting his degree (1551), Cesalpino succeeded Ghini as professor of medicine and director of the botanical garden. He stayed there most of the rest of his life, however, ending his career in Rome as the papal physician and professor at Sapienza.

As a person, Cesalpino remains (to me) a ghost beyond his professional attributes and accomplishments.

In deference to his salaried calling, Cesalpino published miscellaneous descriptive medical papers. For example, he studied the movement of the blood—a whole book (Arcier 1945) tells about this work. Perhaps it was then the best, but reasonable proof of the circulation (rather than a shuttle, back and forth) of the blood awaited Harvey, twenty-five years after Cesalpino's death.

Professor of medicine or not, I suspect that Cesalpino's security in the charts of the angels rests primarily in botany. Probably his interest in plants derived from Ghini and from his subsequent botanical garden responsibilities. Or perhaps earlier he had been a would-be botanist, and these provided the means and the way. So in 1683 he published a book, *De Plantis Libri XVI*. It can be called a book of sixteen parts or it may be considered sixteen books. The first book included his sermon on the mount, the principles of botany, published at a time when there were none such. Then the word, or part of it, was put to work in an arrangement and description of some 1,500 plants known to him. The botany principles, however, were not just of classification as might be inferred from the tail seeming to wag the dog, but included also physiology (primarily plant nutrition), morphogenesis, and terminology. But physiology yet needed rational chemistry, the beginning of which awaited Lavosier yet a couple hundred years off. The nature of plant growth was mostly speculation; for example, Cesalpino had the romantic notion that the juncture of the shoot and the root (Cor) represented the heart of the plant. The text is marred (for us at least) by other attempts to draw analogies between the organs of plants and animals. On the other hand, as **Aristotle** was before him for animals, Cesalpino evidently was an avid observer and presented lucid descriptive interpretations of plant parts, their modifications, and functions. However, it is the systematics for which he is now primarily remembered, and I then limit these paragraphs to that topic and its overtones.

One book was devoted to seedless plants. Cesalpino described more fungi than any predecessor. These naturally were mostly the large fleshy types. Then he climbed the phylogenetic ladder to the ferns and their relatives. The remainder were of seed plants. These

Andrea Cesalpino

Cesalpino divided into herbs, shrubby herbs, shrubs, and trees. This seemingly was just following Theophrastus, and it does appear that he might have made some improvement over the Greek seer. Then Cesalpino turned to generation (reproduction), where he believed the fruit, the seed, and the seedling to be most important. Of fruits he observed, for example, the number of locules, sometimes placentation, and the difference between a superior and inferior ovary. Vegetative characters were felt to be poor differentia for major groups but could be used for the ultimate subordinate "genera." Only limited attention was given to the diverse features of flowers. This may be because of Cesalpino's fixation with the fruiting generative issue. And perhaps he lacked a good hand lens.

Pretty crude, you say, and I can't quarrel. But in line with the old tritism that the proof of the pudding is in the eating, Cesalpino's groupings within the awkward outer shell of herbs, shrubs, etc., were surprisingly natural as we presently conceive them. I mean that many plant families, for example, grasses, umbellifers, composites, mints, etc. (Morton 1981, gives some twenty examples), were treated as middle level taxa with relatively few misnomers. In defining these groups (families to us), Cesalpino seemingly did not use any initially chosen feature or essence. He grouped individual kinds by intuition on the basis of multiple character association, then decided *a posteriori* what the characters were after assembling the group. Character association was not otherwise explicitly espoused for nearly 200 years until by **Adanson** in 1760.

It is repeatedly stated, albeit with minimal documentation, that Cesalpino was a disciple of Aristotle—perhaps authors have sequentially copied from one another. The best appraisal of the relationship of Aristotle and Cesalpino to botany that I have seen is by Morton (1981) and Mayr (1982). But the appraisal of these authors differs sufficiently that one wonders if the authors are talking about the same person.

As I understand the jargon of Aristotelian Scholasticism that might be applicable, it is the concept of essentialism, that of an abstract essence that is determined by intellectual intuition. One classifies not things but their essences. But the essence once established is unitary. There's the problem because it seems that

Cesalpino could be considered an empiricist who used inductive (other authors say deductive) reasoning in his classifications. Perhaps one can partly dodge the problem by assuming that Cesalpino's essence was the fructifications, and that the subordinate categories represent the manifestations thereof.

Cesalpino, of course, knew Aristotle and Theophrastus. Certainly likewise he inherited some notions of the herbalists. So it is with all of us; we start with our philosophical heritage. Conditioned then by the cultural climate of our time, we sieve this through the matrices of our little gray cells, only then asserting our own originality, if any. It is inevitable that Cesalpino obtained inspiration and ideas from Aristotle. But he emphatically was his own man, and went his own way independent and beyond that of the Greek scholar.

So. Cesalpino was the first important elementary or fundamental botanist following Theophrastus. He provided the idea of an interstate highway for those following. Others may subsequently have used different routes and provided better paving. But the initial and eloquent inspiration came from Cesalpino. He is remembered in the leguminous genus *Caesalpinia*, the subfamily Caesalpinioideae, or as some prefer, the family Caesalpiniaceae.

SELECTED BIBLIOGRAPHY

Arcier, C.P. 1945. The Circulation of the Blood and Andrea Cesalpino. New York (Non vidi).
Mägdefrau, K. 1971. Cesalpino. In: Dictionary of Scientific Biography. Vol. 2, pp. 80-81, with citations.
Mayr, E. 1982. Cesalpino. In: The Growth of Biological Thought, pp. 158-161. Belnap Press, Cambridge, Mass.
Morton, A.G. 1981. Cesalpino. In: History of Botanical Science, pp. 128-144, 157-161. Academic Press, London.
Stafleu, F.A. and R.S. Cowan. 1976. Cesalpino. In: Taxonomic Literature. Vol. 1, p. 478, with citations.

Clusius (Charles de L'Écluse, or Carlus Clusius or Jules-Charles L'Écluse) (1526–1607)

Non-cat people find cats monotonously the same. But those who have lived with a few cats for a spell find that each has its own personality and way of "talking" to or about the world. Similarly it may seem to you that those of this sample of botanists (and semibotanists) from the renaissance period are humdrum repetitive. But no. Each had distinctive professional attributes and a special brand of contributions for the future. That of Clusius was his wide-scope flora writing, including cultivated, as well as native species. And beyond his penmanship, Clusius was responsible for the introduction of the common ornamentals of the Near East to Western Europe.

Clusius came of a well-to-do Protestant Dutch family. His formal education and training were for the legal profession (1548). Next I find that he was collecting plants for Rondelet, a physician and botanist at Montpellier. I don't know the causation of this seemingly random quantum jump. Anyway, after three years with Rondelet, he was a botanist. Possessing multilingual skills, he then began translating (and editing and revising where the spirit moved him) botanical (and other) books into Latin. He led a difficult life in which religious persecution included confiscation of all of his (and his father's) financial holdings and property. He fled to

Austria where he lived for a decade or so, part of the time in penury. There he is said to have "found happiness primarily in his inexhaustible capacity for work." It was not until his latter years that he achieved relative security when (1593) he succeeded **Dodoens** in the chair of botany at Leiden. There he planned and established the later famed botanical garden.

As of most botanists of this era, I know nothing of Clusius' private life. Everyone says that he had poor health, but his botanical exploration and the volume of his writing suggests that he was not too seriously incapacitated. Morton (1981) has labeled him as the "most brilliant of the Netherlands botanists." He is said to have been amiable, "charming and a man of few words." (But are not words a usual component of "charming"?)

Now what did he do? I have already mentioned the translations. After a few years in Spain and ten or more years in Austria, he wrote a "flora" of each of those countries! These were scarcely floras in the current idiom; rather they were primarily descriptions of species he knew and new ones he found. His interests were at the species level; he wrote good descriptions but scarcely tinkered with any kind of classification. He is said to have described about 500 new species, a considerable number of them being alpine kinds. He supervised the preparation of illustrations. After his return to Leiden, these and other works were brought together in the book, *Rariorum Plantarum Historia,* his most cited publication. That it yet has reference usefulness is indicated by recent facsimile reprinting. An appendix in this book is a *Fungorum Historia* in which more than 100 fungi are described. One author suggests that he was the founder of descriptive mycology.

Clusius was the first who went beyond native plants to write about those grown as ornamentals. These came from the Ottoman empire that bordered on the Austrian consortium, and its ornamentals included many plants (e.g., tulips, peonies, hyacinths) not known in Western Europe. And Clusius did not just write. Because of his ability to make friends everywhere, he was able to get propagating material of most of these, which he then introduced into western horticulture. The establishment of the Netherlands as the tulip center of the world evidently leads directly back to him. And facing the other direction, he introduced the potato from the

Clusius

Americas into western Europe. All of this led to a European horticultural explosion, and the potato became a staple.

At the time of Clusius' death, Leiden, where he finished his life, had become the botanical center of Europe. So one author states. If so, much credit must go to the ostensibly sickly botanist.

SELECTED BIBLIOGRAPHY

Arber, A. 1986. Clusius. In: Herbals, p. 350, index page references.
Jovet, P. and J.C. Mallet. 1973. Clusius. In: Dictionary of Scientific Biography. Vol. 8, pp. 120-121, with citations
.Morton, A.G., 1981, Clusius. In: History of Botanical Science, p.144. Academic Press, London.
Stafleu, F. 1967. Carolus Clusius' Austrian Flora. Taxon 16:535-537.

John Gerard (1545–1612)

John Gerard (*Gerardia* L., Scrophulariaceae) wrote the most popular of herbals. It undoubtedly exceeds any other in total number of copies printed because it has been reproduced in the twentieth century. It tantalizes the historians both in substance and concerning the nature of its composition.

Gerard came from the presumably wealthy English aristocracy. Nevertheless while yet an adolescent, he was apprenticed to a barber-surgeon for seven years (maybe his parents wanted to get rid of him). He then traveled for several years as a ship's surgeon. This out of his system, he became sedentary in an affluent area in London. And he planted a garden. Maybe this needs explanation. There were no Rolls-Royces available at this time. An alternative way of showing that you were of a higher class than the Duke of Jones was to assemble a fancy garden. Usually a horticultural caretaker was employed to do the work. Gerard, however, was not only his own horticulturist but a successful procurer. He knew both Raleigh and Drake, and the thousand species he acquired were rich in New World kinds, then the rage. This he advertised by publishing a checklist of his wonders. He also became caretaker of the physic garden (medical plants of the College of Physicians of London). He acquired a formidable botanical and horticultural reputation.

John Gerard

Of Gerard's personal life, nothing seen. I skip his political activities among the barber-surgeons, which was perhaps the American Medical Association equivalent of its time.

It is Gerard's *Herball* or *General Historie of Plants* from which his fame (or infamy) derives. Its composition is variously interpreted. Summarized from **Arber** (1986) and Anderson (1977), it may have been about as follows. The final work of the Flemish botanist **Rembert Dodoens** was published in 1583 in Latin and was well received. Although there was already an English edition by Lyte, the Queen's printer, John Norton thought an English edition might have market value. He commissioned a Dr. Priest to prepare a translation. Priest died before this was finished. Norton prevailed on Gerard to finish the job. Gerard, by account, finished the translation, altered the arrangement of the manuscript, and blended it with material from a book of a contemporary herbalist, l'Obel. Then l'Obel became part of the action, taking the responsibility for matching preempted woodcut blocks for illustrations with Gerard's (?) text. But Gerard pushed the work into publication before l'Obel finished his part. Which illustration went with which description was yet a mess. The book came out (1636) as Gerard's *Herball*, the author(?) stating that Priest's translation had disappeared with his death and that he, Gerard, had written a new book to take its place. Beyond the textual-illustration sloppiness, the condemning charge against Gerard, per Arber (1986) and Anderson (1977), was "obvious plagiarism." On the other hand, William Stearn of the British Museum, most commonly a prosecutor of others, recently came to the defense saying that Gerard probably had a book in preparation before getting involved with the Dodoens-Priest-l'Obel affair and that the "Herbal as published contains so much that undoubtedly came from Gerard himself . . . and was so massive a task, that it seems charitable to credit him with the whole" (Stearn 1972).

Whatever the parentage and pregnancy of the herbal, it seems agreed that its unique style makes it more than an ordinary herbal, which has carried it to the present. It is an Alice-in-Wonderland window into the Elizabethan era in England, and is yet studied by historians who may have little interest in botany. A slight taste from Gerard's preface: "Now with our friendly labors we will accompany thee, and lead thee through a grass plot (an account of the

Graminaeae). Then little by little thee through most pleasant gardens and other delightful places where any herb or plant may be found, fit for meate or medicine."

My supposition about all of this is that maybe Gerard did swipe from the work of others, but the writing could only have been his.

A new edition of Gerard published after his death had an editor (Thomas Johnson) who corrected the numerous misdemeanors of the first. It is commonly cited as "Ger. emac" (Gerard emaculatus, meaning Gerard cleaned up). Fine but beside the point; it was plainly the style that immortalized it above others. Gerard would have flourished today. In fact he does. The *Herball* is advertised as a reprint edition by Dover Press, 1990, 1,678 pp., $75.

SELECTED BIBLIOGRAPHY

Anderson, F.J. 1977. The Herbal of John Gerard. An Illustrated History of the Herbals, pp. 218-226. Columbia University Press, N.Y.
Arber, A. 1988. Gerard. In: Herbals, Their Origin and Evolution, 3d ed., pp. 129-132.
    A reprint of the 2d ed (1938) with biographical annotations by W.T. Stearn.
Stearn, W.T. 1972. J. Gerard. In: Dictionary of Scientific Biography. Vol. 5, pp. 361-363.

Gaspard (or Caspar, or Kaspar, or Casparus) Bauhin (1560–1624)

Gaspard Bauhin represents both the zenith and the termination of the herbal period. He was a Swiss botanist and physician who came from a family of similar inclinations. Gaspard and his older brother Jean were both botanists and doctors, their father was a physician, and so was Gaspard Jean, the son of Gaspard here.

Gaspard Bauhin's university and postgraduate work, 1572-1580 was primarily in medicine. (Per dating of Whitteridge [1970] he was in university at the age of 12). Immediately on getting his M.D. at the University of Basel (1580), he was appointed to the faculty. In 1582 he became professor of Greek; in 1589, professor of both human anatomy and botany; he subsequently became dean of the faculty and four times rector of the university. When possible, he balanced responsibilities by being a doctor during the winter and botanizing with students in the summer. In total, his work and publication in human anatomy and physiology probably overshadow his efforts in botany, but that is not relevant to the account here, except to wonder how he did it all. He is treated by Whitteridge (1970) primarily as a physician; the paragraphs describing his botanical work contain ambiguities.

Gaspard Bauhin is said "to have been a delicate and backward child." If so, he recovered. Also, thrice married, with the cooperation of wives a total of four children were produced.

Among Bauhin's several major publications, including a new edition of **Mattioli's** *Commentarii*, it is his terminal *Pinax Theatri Botanici* (1623) that constitutes a landmark and continuing reference in botanical history (several posthumous editions prepared by others followed the original). The *Pinax* classifies the plant kingdom and lists some 6,000 species. If the structure of his seventy-two sections are taken to represent Bauhin's taxonomy above the generic level, it is a sad effort. Here he used the antiquated and threadbare division into trees, shrubs, and herbs, intermixed with sections based on habit, specific features of gross morphology, or utilization, for example, the *Aromata* is about the spices. On the positive side, most of the species of some modern families are listed together; for example, grasses, mints, umbellifers, and the legumes in part. This was possible because his grouping of genera was based on combinations of characters, not any single one, such as the fructification of others that provided the holy doctrine.

Thus, it is at the generic and specific level that Bauhin's light shines brightly. His thesis was that the genus and species are the fundamental units of nature, and he gave it authority. Many of his generic names, later picked up by **Linnaeus,** are in use today, and, on the whole, the generic concepts are substantively good. For species differentiation within each genus, Bauhin used the prevalent diagnostic polynomial phrase that he diligently whittled down to a concise minimum, commonly to one to four descriptive words. The diagnosis and the name were the same thing. (And so it yet remains today with the chemists, e.g. 2,4-dichlorophenoxy-acetic acid is both name and diagnosis and must suffer the nickname, 2,4-D.) Indeed Bauhin was able, especially in small genera, to reduce his diagnosis to one descriptive word following the generic name, this resulting in a combination of only two words (generic and specific). Consequently, he is sometimes credited with the introduction of binary nomenclature. That is incorrect. The subjects were just reduced polynomials that were used only where feasible. Only with Linnaeus came the consistent binomial and a separation between the specific epithet and the diagnosis.

Gaspard Bauhin

Despite the chaotic upper-level classification, Bauhin's relatively coherent genera made the *Pinax* an invaluable reference work. But it went further. To provide a guide to all previous herbal knowledge, the names in prior works were referenced to the species he described. (Linnaeus, who classified everything, including botanists, later labeled Bauhin as a "synonymist.") Bauhin held reasonably that all plants should have just one name, presumably his. All of this, the product certainly of much industry, provides the *Pinax* major stature as a fundamental pre-Linnaean concordance.

The early herbals contained much superstition, for example, the doctrine of signatures, astrological speculations, and mysticism. Bauhin and most of his contemporaries dumped this in the ash can. True, the medical virtues of plants were mostly of folklore derivation, but the alleged efficacy was based on natural premises. Another major advance in Bauhin's procedure was that he accumulated a herbarium of about 4,000 specimens (it is still in existence), and he traded specimens with other workers as a means of assuring himself that they were talking about the same plant. But I must note that Bauhin did not invent the herbarium; the idea of preserving plant specimens by drying them between pieces of paper probably originated with the Italian botanist **Ghini** in the early part of the 1500s.

Reiterating my introductory statement, Bauhin was the epitome both of the culmination and conclusion of the herbal period. What had been the herbals now split into two kinds of productions, the strictly medical pharmocopoeia, on one hand, and the botanical flora (such as the subsequent productions of **Ray** and **Tournefort**) on the other. In the latter, now unfettered by what cures what, a more sophisticated classification of the living world was a major objective.

William Harvey (of blood circulation) called Bauhin "a rare industrious man." Linnaeus named the legume genus *Bauhinia* for the Bauhin brothers, Gaspard and Jean. Exceptionally fitting: *Bauhinia* usually has two-lobed leaves.

SELECTED BIBLIOGRAPHY

Arber, A. 1938. Bauhin. In: Herbals, Their origin and evolution, pp. 114-116. Reprinted

with annotations, 1986. Cambridge University Press, Cambridge.
Stafleu, F.A. and R.C. Cowan. 1976. Bauhin, Caspar. In: Toxonomic Literature. Vol. 1, pp. 147-149, with citations.
Whitteridge, G. 1970. Bauhin, Gaspard. In: Dictionary of Scientific Biography. Vol. 1, pp. 522-525, with citations.

Jan (or Joannes, Joan, or Johannes) Baptista Van Helmont (1577–1644)

A comprehension of how life operates lagged far behind descriptive knowledge of the common animals and plants. Animals as you and I, or snakes and beetles, need food that we get by eating plants or other animals. We also obviously need air and water to survive. Plants are harder to figure out. As they don't eat, seemingly the soil, extracted by the roots and somehow digested, is their food. They need water—obvious from periods of drought—but, unlike animals, apparently not air. Green plants, however, need light. How and why? One can only turn to the joys of fanciful speculation.

If you are a botany student, you likely have heard of Van Helmont because he is commonly mentioned in introductory botany texts. And though I already knew of him, it was not until recent reading prior to preparing this essay that I realized that plant nutrition was just one minor facet of his research and writing. In total, he was an odd combination of a pioneer in the early fetal stirrings of chemistry, at times an alchemist (a seeker for the transmutation of base metals into gold), one who dabbled in both naturalistic causes and mysticism, one who wrote about human physiology and the nature of diseases, the one who discovered carbon dioxide and presented a theory of gases, and—you name it and it is probably there. Asimov (1971) remarks, "he tried to fuse chemistry and religion into something that was not quite either."

Let's back up a mite. Van Helmont was a Flemish (Belgium) physician. You can take your pick among the various transliterations of his first name. He went to school, eventually ending up with an M.D. (1599) and dissatisfaction with everything he had learned. He married Margerite van Ranst and "through her became manorial lord of Merode, Royenborch, Oorschot, and Pellines" (Pagel 1972). In other words he probably subsequently had enough cash to do whatever he damned please. The marriage also resulted in "several" daughters and a son, who in addition to being a "wandering courtier and scholar," edited his father's assembled writings.

Of primary interest to us, Van Helmont seems to have been the first to use experimental quantitative methods to study a biological problem, specifically plant nutrition. From his *Ortus Medicinae,* pp. 108-109 he said (in translation obviously), "That all vegetable [matter] arises from the element of water I learned from this experiment." The experiment consisted of planting a willow shoot weighing 5 lbs. in a container holding 200 lbs. of earth, dry weight. He watered his willow faithfully for five years, at the end of which time it weighted 169 lbs. (plus the weight of the leaves lost each season that he did not compute), while the soil lost only 2 oz. in weight. "Therefore 164 lbs. of wood, bark, and root have arisen from the water alone." These results and conclusions were later validated in a slightly more sophisticated manner by both Robert Boyle (chemist) and **John Ray** (botanist).

How come Van Helmont didn't consider the air? After all much of his chemistry was about gases, most specifically the air, and he concluded that the air was not just one entity but a mixture. This he demonstrated at least with respect to the unique qualities of carbon dioxide—"spirit of wood" as he called it, and the word *gas* comes from him by circumstance. He said that gases constituted "matter in chaos" or some such. He used the Flemish word for *chaos,* of which the phonetic spelling came out as *gas.* Lavosier picked up gas about 150 years later and so it has been since then.

But why again did Van Helmont, a true pioneer in studying the nature of the air, and the discoverer of carbon dioxide, fail to consider the possibility that air was somehow involved? And thus why does this willow (*Salix*) monkey business deserve any historical notice since all it did was mislead everybody for 100 years? To the first, one can only say that his discernment (or lack of it to our privileged eyes) might have been the fact that we often overlook the obvious, as well perhaps as being symbolic of the level of understanding of biological phenomena of the times. On the positive side, while others limited themselves to talking of the theoretical wonders of "the experiment," he did one, the first to attempt to solve the mystery of from what do plants grow. And he did demonstrate that the soil was not the primary food. Others eventually went from there.

SELECTED BIBLIOGRAPHY

Asimov, I. 1971. Jan Baptista Van Helmont. In: Asimov's Biographical Encyclopedia of Science and Technology, pp. 102-103. Doubleday & Co., Garden City, N.J.
Hanneway, O. 1984. J. B. Van Helmont. Amer. Sci. 62:313. Review of Pagel, W. 1982. Joan Baptista Van Helmont, Reformer of Science and Medicine, 219 pp. Cambridge University Press, Cambridge.
Harvey, R.B. 1929. Joannes Baptista van Helmont. Plant Phys. 4:542-546.
Morton, A.G. 1981. J.B. Van Helmont. In: History of Botanical Science, pp. 176-177, 223-224. Academic Press, London.
Pagel, W. 1972. J. B. Van Helmont. In: Dictionary of Scientific Biography. Vol. 6, pp. 253-259, with citations.

John Ray (1623-1705)

Ray (originally Wray) was a natural philosopher in the broadest sense. Among botanists, he presently seems but little known except among historically minded systematists. However, Morton (1981) in a history of botany devotes twenty pages to him and (p. 195) says that he influenced both the theory and the practice of botany more decisively than any other in the last half of the seventeenth century. But *Sic transit glori mundi.*

Born to the working class, his father the village blacksmith, Ray went to Trinity College, Cambridge, with the financial help from a vicar friend (B.A. 1648, M.A. 1651). At Trinity College he subsequently lectured—sequentially greek lecturer, mathematical lecturer and tutor, humanities lecturer, praelector, junior dean, steward (Raven 1950, pp. xv–xvi). He seemingly had before him a made-to-order university and church career. But a conscientious objector of his time (1662), he refused to sign a loyalty pledge (the Act of Uniformity) to the crown and was out on his ears. Most of the rest of his life, he and his family were precariously supported by his wealthy companion Francis Willughby. Freed from other obligations, he devoted himself entirely to his calling, natural philosophy.

Ray married in 1673. There were four girls. Additionally, fol-

John Ray

lowing the early death of Willughby, Ray and his wife became responsible for the education of Willughby's two sons (Miall, 1912, p. 102). Ray is said to have had poor health, but this is also stated for others of outstanding production, such as **Darwin**. Perhaps unable to play football, ski, or climb mountains, these individuals tended to concentrate on the gentle arts of thought and writing.

Although Ray was much more than a botanist, I must primarily limit this account to plants. While still at Cambridge, Ray wrote a local flora of that region. Since there was no British flora, he started on that. Then often with his pal Willughby, he traveled widely over Europe observing and collecting. The culmination of his botanical publication, the multi-volume *Historia Plantarum,* appeared towards the end of his life. Published 1686-1704, it included a botany introduction and an encyclopedic treatment of ca. 18,600 species. Of the latter, he was preceeded only by **Cesalpino** in the use of associated multiple characters in categorizing groups. He defined species—"each produces only its own kind; one must distinguish between essential, accidental, and environmental characters." Though the genera were of irregular nature, Ray, however, often showed excellent intuition in their delimitation. But higher categories were less satisfactory. Views that the influence of this monstrous endeavor was but temporal probably derive from the fact that it somewhat fell apart by the sheer weight of numbers, and that it was rapidly displaced by the more convenient mechanical system of **Linnaeus.** But it was a jump over Linnaeus, whose arbitrary counting of stamens and pistils was essentially a blind alley. Thus Ray is now seen as the direct progenitor of the avowedly "natural" classification of **de Jussieu's** (1789) that is the ancestor of all subsequent ones.

Space allows neither enumeration of Ray's several experiments in plant physiology nor of his writing of its principles in the first volume of his *Historia.* These, however, bring from Morton (1981, p. 210) the probably dubious assertion "Ray can well be regarded as the founder of plant physiology." And Ray's accomplishments in total lead his admirer, Morton (1981, p. 211), to say that Ray was "the consummate botanist of his age." There are, of course, dissenters; Miall (1912, p. 124) views the *Historia* as "a vast compilation . . . proof no doubt of Ray's industry and candour, but little memorable on other grounds."

Beyond botany, Ray wrote of birds, mammals, fishes, insects, geology, and fossils. He produced the first quasi-modern classification of animals. Of birds, I have an Alfred Newton quotation (Raven 1950, p. 308): "The foundation of scientific Ornithology was laid by the joint labors of Francis Willughby and John Ray." A deeply religious man, he came to grips with the philosophical problems of God and nature in his *Wisdom of God* (1691). Therein, he rejected the Aristotlelean thesis of a world co-eternal with an impersonal God, the random roulette wheel of the Epicurean atoms, and Cartesian mechanism. Rather (if I understand correctly) his was a dualistic approach that maintained that while our sense mechanisms are the only means of interpreting reality, there must be "some intelligent plastic Nature" that presides over it all. This book, going through numerous editions, was Ray's bestseller.

Although views about Ray's stature and influence may differ, it is plain that he was a major figure in the evolution of plant classification, and that he was one of the most broadly based and productive botanists of all time. For a presentation of the total Ray one must turn to Charles Raven's *John Ray* (1950), a labor of love and dedicated "To all who, like John Ray, have sacrificed security and careers for conscience sake." (Raven was a controversial pacifist during war years.) Yes Ray did sacrifice one career. But he also had the guts to fabricate another in which he won distinction.

Eponymy that has followed distinguished botanists has usually been in the form of naming genera and species for them. So it is with Ray, for example, the genera *Rajania* (plus orthographic variants of other authors) of the Dioscoreaceae and *Janraia*, also Dioscoreaceae. But more exceptionally, The Ray Society. Founded in London in 1844, its "first rule" has been "The promotion of Natural History by the printing of original works . . . in Zoology and Botany; or new editions of works of established merit . . . and of reprints . . . of works which are generally inaccessible." The society still exists and has been faithful to its pledge for nigh 150 years now. A recent major reprinting known to some current botanists is a facsimile of Linnaeus' *Species Plantarum*.

SELECTED BIBLIOGRAPHY

Miall, L.C. 1912. Ray and Willughby. In: The Early Naturalists, pp. 99-130. MacMillan Company, London.

Morton, A.G. 1981. Ray. In: History of Botanical Science, pp. 195-214, 228-231. Cambridge University Press, Cambridge.

Raven, C.E. 1950. John Ray, Naturalist, His Life and Works. 2d ed., 506 pp. Cambridge University Press, Cambridge.

Webster, C. 1975. J. Ray. Dictionary of Scientific Biography. Vol. 11, pp. 313-318, with citations.

Anton (or Antony) Leeuwenhoek
(or van Leeuwenhoeck) (1632-1723)

Leeuwenhoek was not a botanist; indeed he was not an educated scientist in any sense. He was rather, as Columbus, one who discovered a new world, this a nether realm of microscopic life that, within the limits of his allotted years, he eagerly explored. It was a world that doubled the scope of all biology. For botany, Leeuwenhoek was primarily a prophet.

Developments in biology and other sciences are commonly linked with advances in instrumentation, and Leeuwenhoek is sometimes called the father of the microscope. This is not correct. The microscope, like the telescope a combination of two lens, had been around somewhat before his time. And Leeuwenhoek did not even use a microscope. Rather he employed single, tiny lenses, honed with incredible skill, which were far superior in magnifying resolution than any other equipment then existent. This accomplished, he did not sit on his anonymous laurels; rather he enthusiastically looked at tiny things (e.g., protozoa, bacteria, pollen, fungus spores) and wrote about them. Forms of life that no one had ever seen or perhaps even dreamed of before!

As then a prophet, and a candid though unsophisticated philosopher, he was an oddball among the luminaries of biology.

Anton Leeuwenhoek

We know more about him than most others of his era because of the research of Clifford Dobell, who wrote a marvelous Leeuwenhoek book (Dobell 1932).

Leeuwenhoek was born in the Dutch town of Delft of a commercial family. His formal education was negligible. At the age of sixteen he went to Amsterdam, worked in a linen draper shop and learned the business. After about six years he returned to Delft, set up shop as a draper and haberdasher, and there remained for the rest of his long life. He was a minor city official. As chamberlain to the Sheriffs of Delft, it was evidently his duty to keep their chambers neat and shipshape and light candles when they wanted them and the like. He had a job as city surveyor and was also *winegauger* for the town. He was a friend of the later famous painter Jan Vermeer who was also of Delft, and was responsible, following Vermeer's early death, for the legal proceedings in settling the latter's negative estate and the affairs of the widow and eight children. Leeuwenhoek himself married twice and outlived both wives by some years. Of six children, all died in infancy save one, a daughter Maria, who never married and instead stayed with him his entire life and cared for him in his old age.

All of this seems sufficient to keep a man reasonably preoccupied, but he was apparently doing something else while the other worthies of Delft occupied their free time with whatever was then the equivalent of watching TV. Evidently he spent some years developing a precision skill in producing lenses, which were unparalleled perhaps for more than a 100 years beyond. A lens that is yet extant magnifies about 270×; some of his observations suggest that he used lenses of perhaps 500× resolution. It has been guessed that since drapers commonly used low-magnification lenses to examine the quality of cloths, that his interest in lenses was originally kindled by such routine employment. This may be so, and he went from there. In any event, while he was secretive about his methods of production, he was willing, even eager, to write about the unheard of marvels he had found. And beginning about the age of forty he so continued for nearly half a century. It made him temporarily one of the best known if not most famous biologists of all time—royalty and others great and small from all over Europe flocked to see him and look through his lens.

Famous! And this was Leeuwenhoek, uneducated, except for his own limited reading, who knew no language save a labored colloquial Dutch, who rarely left his city gates and who published only by writing letters.

How did this come about? The Royal Society of London at that time had an extraordinary secretary, Henry Oldenburg, who attempted, on the behalf of the society, to establish contact with all of the "philosophers" of his day. A couple of individuals who had witnessed some of Leeuwenhoek's wonders told Oldenburg about him, and the latter promptly wrote essentially, "Please tell us what you have got, fellow?" Leeuwenhoek reluctantly answered in part as follows: "I have been oft-times besought by diverse gentlemen to set down on paper what I have beheld but I have declined . . . because I have not been brought up to languages . . . and because I do not gladly suffer contradiction or censure from others." But his letter did continue with some illustrated observations about the structure of a bee and a "memoir on what I have noticed about mold." This attracted sufficient attention that he was invited to write more. Which he did. And some of it seemed so incredible, particularly his description of an unknown world of little animals, protozoa, that the society sent **Robert Hooke** (the microscopist who drew cork cells that show up in nearly every beginning botany text) to be sure they weren't being fed a line. Though with inferior equipment and observational techniques, Hooke validated enough to assure that Leeuwenhoek was for real. The Royal Society eagerly solicited more and more, which they received for about fifty years. They elected Leeuwenhoek a fellow of the Royal Society in 1680—"a full fellow and not a Foreign Member"—this even though he never attended a meeting.

What did Leeuwenhoek see? His special darlings were his little animals, protozoa and bacteria, both from water of diverse kinds, and from the insides of various animals including his own sputum and feces. He observed blood cells moving in the tail of a tadpole, this, with Malphighi, validating Harvey's theoretical capillaries. He saw tiny motile animals in the male genital fluid of several animals and developed his own spermaticism theory. "I know there are whole universities that won't believe there are living creatures in

Anton Leeuwenhoek

the male seed; but such things don't worry me because, I know I'm in the right."

Leeuwenhoek wisely was an empiricist. He told about what he saw and mostly eschewed speculation, which he regarded as an amusement of those in academia. But nevertheless, his observations did lead to some "taking sides" in the rife discussions of the time. As noted above, his spermaticist (or animalicist) thesis of reproduction was in opposition to the ovist proposition (as of Harvey) that asserted that the egg did everything, the seminal fluid supplying only stimulative vapors. Leeuwenhoek also denied spontaneous generation—that worms, mites, and other vermin, even mice, originated from the breakdown of organic matter. He saw, instead, that these creatures were stages in the life cycles of various insects, fleas, lice, and flies.

I repeat that direct botanical observations by Leeuwenhoek were minor, and that he was perhaps primarily a prophet. But such observations did include miscellaneous items as some blue-green and green algae, a few "molds," wood structure with pitted tubes, and the role of seeds in plant reproduction. It is possible that other botanical observations have been obscured because his biographer Dobell was a protozoologist and did not recognize or know what to make of them.

For example, Dobell says, "there follow some observations of whose precise significance I am uncertain." Okay, uncertain. But now look at a few sentences excerpted from a page and a half of Dobell's Leeuwenhoek (1932, p. 181-182).

> People are stricken with fever because their shoes get very red whenever they walk through the grass and . . . [they] conclude that the air is infected . . . I turned my attention to the grass itself and saw that some of it was studded with reddish dots. Bringing these before my . . . [lens] I saw that they consisted of small globules where of a thousand did not equal the bigness of a sand grain. They came not out of the air but of the grass itself, and whenever one's foot strikes against the grass, the said globules are dusted off it, and makes one's shoes reddish.

Well, probably some of you can guess even from this fragment of Leeuwenhoek's commentary that he was looking at the uredospores of a rust fungus. They come from the grass, not the air. That is all. But it was well nigh 150 years before anyone went further.

SELECTED BIBLIOGRAPHY

Dobell, C. 1932. Antony van Leeuwenhoek and his "Little Animals," 435 pp. Dover Publication, N.Y. (Reprint 1960).
Heniger, J. 1973. A. Leeuwenhoek. In: Dictionary of Scientific Biography. Vol. 8, pp. 126-130, with citations.
Mann, A.L. and C.C. Vivian. 1963. Anton van Leeuwenhoek. In: Famous Biologists, pp. 41-49. Museum Press, London.

Robert Hooke (1635–1702)

Hooke (English) has been called a physicist, a microscopist, a tinkerer, and more. He was all of these, though scarcely a botanist. But he is included here because a drawing of his, a microscopic cross section of cork, continues to be reproduced in numerous beginning botany and biology texts with the message that Hooke was the first to see and name cells as cells. Yes, he saw, and perhaps that justifies his textbook immortality, even though his idea of what cells are was way off base.

At an early age, Hooke showed spectacular talents in mechanics and had artistic abilities. He studied literature, languages, and mathematics in secondary schools, and read about everything else. He obtained an M.A. at Oxford. He was a member of a high-toned snob group that subsequently became the Royal Society of London. There, he obtained employment (of a kind) and became the scientific idea person: "Curator of Experiments." He had an irregular small salary and lodging at Gresham College where he lived for the rest of his life, some forty years. And that indeed was his life, inventing, throwing off new ideas, and quarreling with others.

Hooke is described as having a wretched appearance, wretched health, and hypochondria. Seemingly he never married, but Westfall (1972) says his niece, Grace, was his mistress until her death (1687) "through a prolonged and tempestuous romance." That's about all we know about the man.

As Curator of Experiments, Hook developed an air pump for Robert Boyle that was used in the formulation of Boyle's law. He dealt with clockmaking. Clocks had to be large to accommodate the pendulum. So he worked out the mechanical theory for spiral springs that made possible small watches and the chronometer; he was surveyor and architect for rebuilding after the great fire of London; he sparred with the younger super-genius Newton with respect to theories of universal gravitation; he regarded fossils as evidence of long gone life; he presented a theory that light consists of "pulses of motion." He was secretary of the Royal Society following Oldenburg. This list, which could continue for two pages, gives some idea of his versatility. He had ideas on more things than he could follow up, and this led to disputes with most of his competitors.

The Italians had invented the compound microscope and then apparently lost interest. Hooke and a few others followed up; he improved his microscope to the extent that it was probably the best in existence. From this superior vantage point he looked at everything. There was much to write about because no one else had ever seen it (like those who first used the electron microscope).

His *Micrographia* was published in 1665. It was presumably a compilation of everything he had seen. Actually it was quite a bit more. There was a lengthy preface that has little to do with the primary text such as his studies in meterology. When Hooke thought of something he wanted to write down, he did so regardless of context; and he broadened the scope of his readership by writing in English rather than Latin. Hooke examined everything, animate or inanimate (i.e., mineral) he could conveniently get his hands on. The grab bag of biological material included such things as a horrible, frightening picture of a flea and the eye of a fly—he was an excellent artist. He saw that diverse plant tissues were "all perforated much like a honeycomb" and the cork illustration with the text accompanying it supports this statement. But his interpretation of the perforations as "cells" was completely different than the sense in which that word has subsequently been used. Instead they were viewed as a medley of pores through which, he assumed, plant nutrients moved. And he asserted that these were perhaps universal in plants. The name and the idea stuck.

Robert Hooke

Then the great fire of London struck, and Hooke was busy with plans to rebuild the city. That was the end of his botany.

SELECTED BIBLIOGRAPHY

Kelly, G.F. 1965. Robert Hooke plus 300 Years. Bioscience 15:408-411, with citations.
Morton, A.G. 1981. R. Hooke. In: History of Botanical Science, p. 178. Academic Press, London.
Westfall, R.S. 1972. R. Hooke. In: Dictionary of Scientific Biography. Vol. 6, pp. 481–488.

Nehemiah Grew (1641–1712)

Nehemiah Grew, English, was the first microscopist who was entirely a botanist. Only the brilliant Italian, Malpighi, was in contemporaneous production, but Malpighi worked primarily with animal material. Grew's *The Anatomy of Plants* (1682), summing up his earlier publications, was the botanical reference for more than a century. Thus he was both the initial and the continuing patriarch and philosopher of plant anatomy.

Grew's studies at Cambridge were interrupted by war and because his religious views were not correct. He finally obtained his M.D. in Holland and apparently supported himself for most of his life as a practicing physician. In later years as "Curator of Anatomy of Plants" of the Royal Society of London, his salary (50 pounds per year) was paid from donations from "willing members."

Grew married; his wife, Mary Huetson, died; he married again, to a woman named Elizabeth Dodson, and this time there were several children.

The Anatomy of Plants was a pioneering work, both descriptive and idea-productive. Grew's approach was holistic, the entire plant and all of its parts, macroscopic and microscopic, and especially the relation of structure to function. For vascular plants, he found regularity and patterning among a few basic structures and tissues. He said that a plant consists of two types of elements, the fibro-vascu-

Nehemiah Grew

lar and parenchymatous—and the latter word derives from him. While **Hooke** first saw cells and gave them that name, it was Grew (and Malpighi) who found that they were the invariable units of parenchymatous tissue. Grew theorized that this was true also of the vascular tissue, that the vessels (air tubes to him) perhaps originated following the breakdown of walls between linearly arranged cell series. This theory is not quite the same as the much later (1838-1839) **Schleiden** and Schann "cell theory," because Grew, sticking close to what he had seen, did not present cells as the basic structural units of life nor did he theorize about their origin from pre-existing cells. And he worked only with plant material. But it comes close.

Grew provided structural definitions for all plant parts; for example, he listed nearly all of the currently recognized morphological differences between a stem and root. He defined leaves in terms of the nature of their origin from stems and described the vascularization connecting them with the stem. He showed that the "flower" of the Compositae is a compound structure. His examinations of stamens led to the hypothesis that they are the male organs of the plant. Presenting the need for experimentation in all phases of botany, he blended miscellaneous physiological investigations, for example, concerning geotropism, with his largely observational and deductive studies. He denied the biblically derived axiom that plants were created for the benefit of man; rather, he said that while man might obtain nutrient and delight from plants, their organization plainly was for the benefit of the plants, not humans.

Towards the end of his life, Grew became loquacious beyond botany in a *Cosmologia Sacra,* a philosophical treatise on his religious beliefs and their application to living.

Because Grew and Malpighi were contemporaneous, there has been some historical bra-hoo about their relationship, especially of overlapping publication about plant anatomy. A denigration of Grew seems to have been initiated in the nineteenth century by the German botanist Schleiden and later seconded by **Sachs** in his *History of Botany.* Their pitch was that Grew pirated from Malpighi, i.e., took descriptions and ideas first expressed by the latter without credit. Carruthers (1902), citing dates and circumstantial data, dismissed the charge. Morton (1981) has vigorously reiterated

Carruther's position, concluding that although the two men knew and respected each other's work (sometimes coming to similar conclusions), their studies were entirely independent. Each was his own man. Morton observes (1981, p. 179) "Malpighi certainly saw more clearly than Grew in some matters of detail, but Grew developed a more integrated view of plant structure."

The following quotation portrays Grew's spirit:

> For what we obtain of Nature, we must not do it by commanding, but by courting of Her. I mean that wherever Men will go beyond Phansie and Imagination . . . they must Labour, Hope, and Persevere. . . . The Way is long and dark. . . . If but little should be affected, yet to design more can do us no harm; for although a Man shall never be able to hit Stars by shooting at them; yet he shall come much nearer to them than another that throws at apples.

SELECTED BIBLIOGRAPHY

Carruthers, W. 1902. On the Life and Work of Nehemiah Grew. J. Royal Microscopical Society 129:129-141.
Metcalfe, C.R. 1971. N. Grew. In: Dictionary of Scientific Biography. Vol. 5, pp. 534-536, with citations.
Morton, A.G. 1981. Grew. In: History of Botanical Science, pp. 178-195. Academic Press, London.

Joseph Pitton de Tournefort (1656–1708)

Modern statements in textbooks proclaim Tournefort as the genus man. And they are correct. J.P. Tournefort came from a substantial family and was Jesuit-educated in the classics and natural history. He studied at the University of Montpellier, getting his fun by mountain climbing and plant collecting. In 1683 he went to Paris as a substitute lecturer at the Jardin du Roi. Although he had published essentially nothing, he seemingly was already well known through correspondence and numerous botanical friends. He ultimately became *The Professor* at the garden and was in midlife when killed in a traffic accident. He left much unpublished material. Books that appeared during his life were *Schola Botanica* (1689), a summary of lecture notes prepared by one of his students; then the *Éléments de botanique* (1694) that was translated into Latin (revised?) and appeared as *Institutiones rei herbariae* (1700), by which title it is usually known.

Posthumous publications were arranged, I suppose, by his friends. Among these, his nonbotanical *Relation d'un voyage du Levant* (1717), two volumes drawn from correspondence, has had the greatest continuing general interest.

But it is the *Institutiones* that we are interested in. It included Tournefort's principles of classification and descriptions of about 7,000 species in 700-plus genera. Tournefort looked at the whole plant kingdom. He recognized lichens as a special group. His descriptive and interpretative work on fungi was not only a beacon ahead for mycologists to come but also included methods for mushroom culture. The classification of flowering plants started with the time-worn division into trees and herbs. Subordinate categories were more happily invoked, for example, flower characters (e.g., petal-bearing vs. apetalous, polypetalous vs. gamopetalous, regular vs. irregular), this contrasting with **Cesalpino's** preoccupation with fruits and seeds. Modern writers say that it was an artificial classification. Certainly "yes" as concerns trees vs. herbs. Otherwise the case is out of context. All authors, including **Linnaeus,** compromised between the twin goals of grouping together plants that they intuitively felt had affinity, and that of convenience, i.e., providing the student an orderly pathway through the maze of observed diversity. Tournefort did a superior balancing act, and his system was the one most used until superseded by that of Linnaeus.

But this is yet not our focus. Tournefort was the one who first gave the genus a clear-cut status in a scheme of classification. **Bauhin** had previously tentatively tried the genus idea for "clusters of species," but he scarcely carried through. Tournefort did so, and his genera were incisive in concept and lucidly defined. Linnaeus paid him the compliment of taking up many of these intact, or in some instances, combining two or more into a single group. Tournefort demanded more internal coherence of his groups than Linnaeus; they tended to be smaller and more precise. The reason nomenclaturally that these taxa are credited to Linnaeus rather than Tournefort is that priority in nomenclature begins with Linnaeus' *Species Plantarum* of 1753. Tournefort was earlier. His contrasting casual treatment of species perhaps reflects his view that the basic biological units of nature are the genera.

In perhaps a minor vein, the use of the word "herbarium" in the present sense seems to derive from Tournefort; among several prior ad hoc terms the name *Hortus siccus* was perhaps most used.

My initial statement was "Tournefort gave us the genus." And it so remains. Of course, in derogatory tone, one can say "so what?" If Tournefort had never existed, the genus as a basic unit in plant classification soon would have come into existence, probably with Linnaeus a few decades later. But, likewise in modern times. If

Crick-Watson had never been around to announce a hypothesis about the structure of DNA, almost certainly within a year or even less, it would have come from one of the other competing groups. You now no longer hear of these others unless you are familiar with the claims and counterclaims of the nucleic acid literature of the late 1940s and the early 1950s. Such is the way of priority. Tournefort's priority was of a unique nature, because at his time he was seemingly almost alone in the idea. So Tournefort and Crick-Watson live.

SELECTED BIBLIOGRAPHY

Leroy, J.F. 1976. Tournefort. In: Dictionary of Scientific Biography. Vol. 13, 442-444, with citations.

Stafleu, F.A. 1971. Tournefort. In: Linnaeus and the Linnaeans, p. 385, index entries. A. Oosthoek, Utrecht, The Netherlands.

Stafleu, F.A. and R.S. Cowan. 1986. Tournefort. In: Taxonomic Literature. Vol. 6, pp. 412-415, with citations.

Rudolph Jakob Camerarius (Camerer)
(1665–1721)

For all recorded history, sex in plants was vaguely believed, disbelieved, refuted, or ignored. It is not possible, in a brief accounting, to dwell on the several ways in which sexuality or not in plants flitted among the several philosophical theses of pre-Mendelian heredity, or generation, as it was called. Suffice to say both **Grew** and **Ray** believed, on an observational basis, that flowering plants, though usually hermaphroditic (a single individual possesses both sex organs), performed the rite of sex, and that the pollen grains produced by the stamens are the male sperm-equivalent. But their suppositions were based on observation and could scarcely be regarded as proof. It was this chap Rudolph Camerarius who provided the experimental evidence that somewhat broke the vacillating stalemate. Presently, we might regard his work as just tinkering, but it was such that most botanists cried, "Voila, enchanté," and plants henceforth had sex.

Camerarius (German) came of a medical-professional family. He was born in Tübingen and died there. After getting his master's degree, he beat around the bushes in Europe for two years, then returned to home roost. There he soon became a medical professor and director of the Botanical Garden. His "epoch-making" publication on plants and sex was as obscure as that of **Mendel's,** a letter

Rudolph Camerarius

that was then printed in a local *Proceedings* (1694), but unlike Mendel's, it attracted attention.

So he did what? Initially Camerarius investigated the development of the embryo in a legume flower and formed the hypothesis that fertilization, i.e., pollination, the transfer of pollen from the anther to the stigma, was a prerequisite. But this was scarcely ahead of the observations and tentative conclusions of Grew. Camerarius, however, then turned to a mixture of further observation and minor experimentation on a few dioecious (male and female sex organs on different plants) and monoecious (male and female sex organs in different flowers on the same plant) plants. The "female" mulberries, he found, set fruits but no seed unless there is a staminate plant in the vicinity. He grew *Mercurialis* (Euphorbiaceae; dioecious; common name "Boys-and-girls" in current **Gray's Manual**). Evidently these were planted in pots and then moved around, thereby giving the operation the dignity of experimentation. Anyway, the girl plants would set no seed unless the boy plant was also there. And the same for spinach (*Spinacia*). He frustrated seed formation in the monoecious *Ricinus* (castor bean) by removing the staminate flowers before they opened and in corn by cutting off the tassels. While he made no similar experiments with perfect flowers, others in the early 1700s rapidly followed up working with the hermaphrodites by removal of stamens before the stigma and style were ready to go to work.

Linneaus' classifications (1735 and of those following years) were primarily based on counting stamens and tabulating the degree of fusion among them and other flower parts. While the function of the stamens was beside the point, Linneaus made the most of the recently accepted sexual role of flowers in advertising his classification system, the nature of which is given in a fragment following (translated from the Latin, of course).

> *Monoecia.* Husbands live with their wives in the same house, but have different beds (monoecious).
> *Dioecia.* Husbands and wives have different houses (dioecious).
> *Polygamia.* Husbands live with wives and concubines (polygamous).

Polygamia aequalis. Equal polygamy; many marriages with promiscuous intercourse.

A tomato, a plant, and a cat, an animal, are very different in most ways. But by whatever name they are the same in that it is the fusion of two nuclei, bringing together different DNA messages, that is essentially common to all. It is only that the arrangements for this happy union differ.

The exceptions to this generality are primitive organisms such as bacteria that lack organized nuclei. So-called prokaryotes cannot be considered either plants or animals. Even among them there are now hints that occasionally lonely DNAs do come together in fusion.

Camerarius' conclusions remained untarnished.

SELECTED BIBLIOGRAPHY

Mägdefrau, K. Camerarius. 1978. In: Dictionary of Scientific Biography. Vol. 15, Suppl. 1, pp. 67-68.

Morton, A.G. 1981. Camerarius. In: History of Botanical Science, pp. 213-220, 231. Academic Press, London.

Stephen Hales (1677-1761)

"Stephen Hales was probably the most important English scientist of the eighteenth century. . . . He made significant contributions to plant and animal physiology . . . and to chemistry," (Home 1981). I don't know about "the most important British scientist," but indeed and certainly he was the first full-blown, quantitative, and deductive plant physiologist.

Hales' professional bucket overflows in the literature; that of his origins and personal life is negligible. He went to Cambridge. There he stayed for thirteen years, studying divinity, getting both B.A. and M.A. degrees. He continued residence as a fellow, working with lively colleagues who were trying diverse experiments in physics, chemistry, and biology. This was the exciting time when Descartes was being washed away by the Newtonian mathematical and quantitative philosophy. There was no conflict of the dual roles of religion and science. Natural philosophy, a study of the product of the Divine Architect, was indeed a religious pursuit.

On leaving Cambridge, Hales became a parish priest, "Perpetual curate" at Teddington (and he was perpetual for fifty years). By accounts he was a conscientious Man of God.

But he was much more. Guerlac (1972) names his specialties "physiology" and "public health." To scatter-list some of his activities: He worked on animal blood pressure; he tried to figure out how to alleviate kidney and bladder stones; he devised ventilators to be used in mines and on slave ships to reduce the mortality among the unfortunate, enforced inhabitants of those human hells; he proposed a more efficient way of distilling seawater and how to preserve biscuit from the weevil; he was an activist about the uncontrolled gin trade and a prominent member of several groups such as the "Society Promoting Christian Knowledge."

But we must here limit this discourse to Hales' plant physiology. This was mostly published in the first edition (1727) of his *Vegetable Staticks* . . . reduction from a polynomial title. One of his most important experimental accomplishments concerned "sap in vegetables." The current notion at this time was that the movement of sap in plants was analogous to the circulation of blood in animals. Hales, studying the how, why, and when of water (sap) movement in plants, set out to see if this was so. He showed, through precise measurements in a series of experiments, the constant uptake of water by plants and its loss through "perspiration" (transpiration). He successfully demonstrated that more than one force was concerned, i.e., that coming from the roots and that from the leaves. Transpiration, he said, goes more rapidly during the day and when it is warm, more slowly in darkness and under high humidity; it shows a daily periodicity. Root pressure is seasonal; the pull of the leaves is the more important force.

Hales devised several methods of measuring and analyzing plant growth, a simple example being that of sequentially counting the number of squares of graph paper a growing leaf, hour by hour would encompass as it expanded. He also tried to determine the role of air in the functioning of plants. In this he may have been the first, but his insufficient understanding of the components of air scarcely led to more than uncertain frustration. At the least, however, his observations justified the suspicion that the leaves of plants "do imbibe elastick air," and that air might have a role in providing for the food of plants. "And may not light also?" Perspicacious suggestions, but verification awaited the work of others.

Stephen Hales was a brilliant man, a good man, and certainly a busy one. I hope he was also a happy man.

SELECTED BIBLIOGRAPHY

Guerlac, H. 1972. Hales. In: Dictionary of Scientific Biography. Vol. 6, pp. 35-48, with citations.
Home, R.W. 1981. S. Hales. Science 211:271-272. Review of Allan, D.G.C. and R.E. Schofield. 1980. Stephen Hales, Scientist and Philanthropist. Scholar Press, London. 220 pp.
Morton, A.G. 1981. Hales. In: History of Botanical Science, pp. 246-254, 280. Academic Press, London.
Woolhouse, H.W. 1966. S. Hales. The New Phytologist 65:559-560. Review of Clark-Kennedy, A.E. 1929. Stephen Hales, D.D., F.R.S. An Eighteenth Century Biography. Cambridge University Press (reprinted 1965, Tregg Press, Ridgewood, N.J.).

John Bartram
(1699–1777)

William Bartram
(1739–1823)

The American colonies and subsequently the thirteen states, naturally, for a long time, could not compete with the Old World in the advancement of science. The first botanists who wrote about plants of the Americas were European born. The first real native American botanist (i.e., born on stars-and-stripes soil) probably was John Bartram, who was immediately followed by his son, William Bartram. In total historical perspective, both, no doubt, were minor actors. But their exploration in the name of love of nature, especially of plants, and their unsurpassed knowledge, is so charming, even romantic, I cannot skip them.

John Bartram, a Pennsylvania farmer, had no education beyond the country school. He was married twice, had a large family, and was a good farmer and businessman. He devoted a small area on the farm to interesting shrubs and trees. He planted them because he liked to do so. There he became so successful that he was able to leave the farming to his family and travel and collect more new and otherwise novel plants. He established contact with European plant merchants and English gentry, who ordered plants and seeds, and his fun became his business. He traveled and collected all over the colonies, New England south to Florida. He kept journals of some of his travels, and they were circulated in London, one published. European explorers sought him out because he was *the* authority on American plants. Although not a publishing botanist, he was observant and had a retentive memory; one individual stated that not a thousandth of what he knew was ever written down. He was elected to various European philosophical societies, and it has been said that he ranked with Ben Franklin as a country-grown, natural genius.

The farm went to his son, John, who had his father's business acumen. It was just as well because William was of another cut. He had no particular desire to do any work as it is usually defined, and he never married. He was seemingly a dilettante with spacey interests in plants, birds, and drawing. Peattie (1936) says, "He was generally regarded as a bit more than a bit daft." He tried to make a living but failed. "So whimsical," his father said. But his drawings, especially of birds, had brought attention, and he had a sponsored chance to go to Florida and do more. There he was, in Florida, South Carolina, and Georgia for four years. It was during the Revolutionary War; he was an innocent among the Indians who somehow recognized him as a harmless eccentric. He recorded everything he saw—plants, birds, the Indian cultures, thunderstorms. When he returned home, the pioneer University of Pennsylvania offered him a professorship. He declined. But he was of more a publishing turn than his father, because he wrote up his journal notes in book form. *Bartram's Travels,* as they are now called, was published first in 1791, and then translated and reprinted in practically all of the western European countries. It was and is a purple-prosed essay on nature expressed in the form of the European romanticism of the time. William Bartram left a literary treasure that, as Thoreau's *Walden Pond,* will be with us as long as people yet read.

SELECTED BIBLIOGRAPHY

Bell, W.J., Jr. 1970. John Bartram, William Bartram. In: Dictionary of Scientific Biography. Vol. 1, pp. 486-490, with citations.
Peattie, D.C. 1936. The Bartrams. In: Green Laurels, pp. 189-199. Simon & Schuster, New York.
Precup, A.V. 1976. John Bartram: 1699-1777. Bioscience 26(5):359.
Stafleu, F.A. 1969. The Bartram Drawings. Taxon 18:443-445.

George-Louis Le Clerc (Comte de) Buffon (1707-1788)

The eighteenth century prior to the French revolution was a time of supreme privilege for a few and but scanty crumbs for most. But, among those who had everything, it was the time of the enlightenment that brought the birth of biology both in substance and name. The major actors included Buffon, **Adanson, Lamarck,** and Diderot of the mighty 36-volume *Encyclopedia of Science.* Among these, Buffon may have been preeminent. Although his scope was all of natural philosophy, i.e., he was only marginally a botanist, his effect on the cultural climate in which botany operated was such that he cannot be ignored.

George-Louis, born with a silver spoon in his mouth, was one of five children in a family whose intellectual, social, and political orientations were all supported by plenty of francs. He was educated by the Jesuits (1717-1723). For the next fifteen years he was peripatetic, doing diverse things such as working on the improvement of war vessels, consorting with royalty, increasing his fortune, all slightly diluted by a bit of botany, which was a translation of **Stephen Hale**'s *Vegetable Statiks* into French. He was a courtier as well as a philosopher at Versailles, dazzling and quick-witted in speech as well as a writing showpiece. He was appointed *Intendent* (director) of the Jardin du Roi (Louis XV) with handsome remuneration. And for the rest of his life, he administered the garden (doubling its size), maintained his several estates and investments, and wrote on everything. The latter was a self-assigned task, one to bring light to all of the ignoramuses before and about him, and as it proves, the primary reason that we know of him now.

Buffon married a rich Marie Françoise de St. Berlin in 1752; there was one son; his wife died in 1769. Buffon was plainly of an arrogant personality, supremely confident of his superiority over others, but he never lacked the right words for the right people, especially for the king's mistress, the Marquise de Pompadour, who it is alleged, held the appointments of the court at her "pretty finger tips." (Peattie 1936, p. 56).

Buffon was a man of organized routine. His regime for life from now on was summers on his estate, other seasons at the garden in Paris. He was up with the birds in the morning and wrote until noon; the remainder of the day was allotted to his financial and social affairs.

Well, then what did Buffon write? Perhaps one can say natural philosophy in all of its ramifications. His "biggy" was the *Histoire naturelle, générale et particulière,* 15 volumes, 1749-1767, and I limit inquiry here to a sampling from that. First it was necessary to provide a definition of science; what it was and what it was not. There he was seemingly the first to make a clean break between science and metaphysics. In contrast, Newton had kept the deity in the picture as the one responsible for the arrangements of the universe; the humble student was to provide an interpretation for the glory of God. Buffon's cosmology gave no such credit; on the contrary what he saw was natural causes through time. A mathematician, he calculated the age of the earth as 75,000 to 200,000 years (in an unpublished manuscript, he estimated up to 3 million).

Obviously this contrasted with biblical views of 5,000 to 7,000 years. Fossils were pertinent to an understanding of the development of life; their explanation through catastrophism (succeeding creations and subsequent catastrophes as the biblical flood) was rejected. Buffon saw change and motion in life and continuity among organisms. Here and later, he wrote of evolution (he didn't call it that) and had to publish a recantation, which, however, he carefully worded ambiguously.

Life is not just a Cartesian machine, lever-operated by the laws of physics and mathematics. Its sum total is more than that of the parts. But it is of this world; there is no vitalism, divine essence, required.

Although Buffon wrote primarily of animals, most of his propositions apply equally to plants. Most conspicuously he was both contemptuous of and hostile to **Linneaus.** Linneaus believed in the fixity of species. Buffon did not. Buffon felt that all schemes of classification above the species level were human artifices "devised only to comfort our own minds." (Roger 1970, p. 580). Plainly he would have nothing to do with genera or families. He was ambivalent about species. He said, "In nature there are only individuals." But he also provided a definition of species identical with and predating Ernst Mayrs' "biological species" a couple of centuries later.

Enough of a sample. What to make of it? There is no doubt that Buffon pioneered before any other in broad concept, cosmology to biology. This granted unanimously, the diversity of views otherwise are a function of Buffon's supreme confidence in himself and his manner of production. By the latter I mean that his voluminous writing was whelmed and extemporaneous. It is said it was never revised. It changed from one year to the next.

Most of Buffon's expansive generalizations have, I believe, stood the test of time. Where they failed it was because of his preoccupation with the big picture, whereas sometimes it is the patterning of the details that provide a truer concept of the whole. His allergy to any kind of a classification of the living world is inconsistent and hard to understand. Certainly some structuring, Linnaean or not, even artificial and human contrived, an index if nothing else, is necessary for communication. A fish is not an oak tree.

But mostly, Buffon was received with acclamation, the superman, by most of his contemporaries. Mirabeau stated that "Buffon was the greatest man of his century and many others." Cuvier regarded him as a god. Voltaire, though he had previously been at odds with Buffon, said that he was a "second Archimedes" (Peattie 1936, pp. 58-59). But sadly this ran down towards the end. The breadth of his all-knowing pompous style perhaps stimulated counter reaction. Situations placed him in competition with

George-Louis (Comte de) Buffon

Réeaumur, the ant specialist who was quiet, modest, and revised constantly, i.e., the apothesis of Buffon. And then Buffon espoused Needham, who, from England, was upholding spontaneous generation ("flies from dung") against Spallazini who disproved them: flies come from eggs that produce larvae that give rise to the flying adults.

Present view has been perhaps best summed up by Stafleu (1967), "Buffon . . . the difficult, often arrogant leader of the group of biologists at the Jardin du Roi produced a book in which the ideas are so far flung that it still makes fascinating reading." Asimov (1972) is less kind; speaking of the grand design in nature, he remarks "better as popular writing than science . . . too easy generalizations and a tendency to superficiality."

Buffon died, perhaps fortunately, the year before the Bastille. His epitaph perhaps can be his own evaluation, *"Le style c'est l'homme meme."*

SELECTED BIBLIOGRAPHY

Asimov, I. 1972. Asimov's Biographical Encyclopedia of Science and Technology. Rev. ed., pp. 164-165.
Peattie, D.C. 1936. Science at Court—Buffon and Réaumur. In: Green Laurels, pp. 56-77.
Roger, J. 1970. Buffon. In: Dictionary of Scientific Biography. Vol. 2, pp. 576-582, with citations.
Stafleu, F.A. 1967. The Biology of the Enlightenment. Taxon 16:431-435.

Carl (or Carolus) Linnaeus (von Linné)
(1707–1778)

More has been written about **Darwin, Luther Burbank,** and **George Washington Carver** than any others among botanists or semibotanists whom I have encountered. The reason for this is evident—they interacted with the public domain and were known beyond the narrow boundaries of science. Linnaeus, however, who follows Darwin not too distantly in bulk of after-the-event literature, is a horse of a different color. True, in proposing a new system of classification of plants, animals, and minerals (*Systema Naturae*), he was widely known in his day, especially to botanists, because he was first and last a botanist. But this was within the shuttered veil of science; he did not trip over religious dogma as did Darwin, and presumably John Q. Public would have little interest in his *Diadelphia Decandra,* et al. But the facts are that Linnaeus was proclaimed a public hero in his country, Sweden, even in his time. And this has continued to the present, for example, in the form of postage stamps honoring him, the most recent (1978) being an issue of some six Swedish stamps on the occasion of the bicentenary of his death. Granted that 200 years ago, Linnaeus' classification of plants with its accompanying innovations were the talk of the botanical town as chloroplast DNA is today. Though mostly acclaimed, he was also reviled (as by **Buffon**). And

Carl Linnaeus

although Linnaeus remains known to botanists to the present, it is especially because the first edition of his *Species Plantarum* (1753) is the starting date for nomenclature of nearly all plants except the preponderance of algae and fungi. The typification of Linnaean species by procedures specified by the present Code of Nomenclature (plus the views of a cadre of Linnaean specialists) is commonly a tricky business. But why and how does any of this rate postage stamp recognition? I have no adequate answer.

The Swedish government has honored Linnaeus several times, issuing stamps about him in 1939, 1963, and 1978, the latter recognizing the two hundredth anniversary of his death.

But to the man himself. Linnaeus was born in southern Sweden. His father was a rural minister who maintained a flower garden on the parsonage grounds. The young Linnaeus was collecting plants from boyhood on. He went to Latin school and then successively to the Universities of Lund and Upsala, presumptively to work on a medical degree. But he found that medical instruction was essentially defunct in both institutions. However, he acquired botanical sponsors who kept him from starving and who fanned his botanical flames. During these student years, his love of nature, especially plants, changed to zest and the zest to a passion for putting the whole business in order. Almost immediately after making a plant-collecting trip to Lapland (northern Scandinavia and essentially his only field work, ever) in his early twenties, he was planning what he would write, listing some thirteen books. He also wanted to marry and be assured of an adequate income, i.e., a good dowry. He found the right person, Sara Moraeus, daughter of a physician. But poppa said that Linnaeus must get his doctorate before the sacred event came off. Linnaeus decided that the best deal would be at the University of Harderwijk in Holland (a ditty of the time: Harderwijk is a town of trades. They sell kippers, blueberries, and university grades). Equipped with his thesis and manuscripts for several books, he got his degree in about three weeks. But then, instead of dashing madly back to Sara, he became entranced with the goodies that the naturalists in Holland had to offer him, especially the splendid, private botanical garden of George Clifford, mayor of Amsterdam and one of the big Indians of the East India Company. Clifford wanted Linnaeus to classify and "write up" his plants, and Linnaeus stayed three years producing *Hortus Cliffortianus* (1735), presently known to all botanical nomenclators. And he wrote some other books on the side, these being the beginning of an apoplectic deluge that continued throughout his life, and which subsequently produced generations of a special cult of Linnaean biographers, bibliographers, and interpreters.

But Linnaeus eventually came back to Sweden and married his evidently patient bride. After a few rough years, he obtained the professorship (botany, medicine) at Upsala and so remained henceforth, a cabinet botanist, teacher, and writer. No more a traveler; he sent his students to the four corners of the earth to gather new plants for him. Perhaps this was a reasonable survival technique because the exploration-mortality among his "apostles," as he called them, was considerably higher than that of current astronauts.

Linnaeus was plainly a flamboyant teacher, and authors speak of the Master leading droves of adoring students out to botanize. He was a religious fundamentalist, and to him the species was an immutable estrade, the produce of the Deity (God created— Linnaeus described). He was a pragmatic, orderly, bibliographic, and descriptive carnivore. Much has been written about Linnaeus' *Philosophia Botanica* (e.g., Stafleu 1971; Stearn 1955) which I pass by, taking up rather his *Species Plantarum* wherein much of his philosophy is expressed in practice. Except, it seems to me he was scarcely a philosopher in the usual connotation of that word, his

Carl Linnaeus

philosophy mostly being expressed in assertive epigrams that essentially said this is what I am going to do, and this is the way it is because I said so. This, plus the tunnel vision constraints of his religious tenets, soon brought him into contention with some of his more broadly concept-minded contemporaries.

Now most importantly, what did Linnaeus contribute that bears directly on the history of botany beyond his time? His most significant donations follow:

A classification system that worked. No one else, **Ray, Cesalpino,** or **Tournefort,** had come close. The novice could quickly learn to follow the virtuoso Linnaeus in plant determination. Pick up a flower, and using a small lens if necessary, count the stamens, look to see if they are free or fused with each other or with the pistil, and quickly turn to the right part of the book, and find what it is. Or for a flower from a dried specimen, Linnaeus would pop it in his mouth for a moment to wet it, and proceed. Wonderful. Furthermore the system was reasonably natural at the generic level because stamen number is commonly consistent within genera.

Good diagnoses. Linnaeus honed both generic and specific diagnoses to a finer edge than any previous student. Generic descriptions are provided in *Genera Plantarum* (six editions), but are not included in *Species Plantarum*. In the latter, the descriptive phrases are polynomials, providing the essential characters clearly separating the subject from any other species in the genus in the fewest possible words. If there are only a few species in a genus, the matter is simple, commonly just a couple of words. For example, his listing of the genus *Glycyrrhiza* (licorice) includes three species characterized as follows: "leguminibus hirsutis"; "leguminibus echinata"; and "leguminibus glabra." More words were required for the representatives of larger genera. Along with this, Linnaeus had a keen intuitive sense for species delimitations, the majority of his concepts remaining valid for now more than 200 years.

Literature. Linnaeus provided a precise synonymy, exceeding any prior. Sure, subsequent research has shown that he made a fair number of errors, but considering the thousands of plants he dealt with, his organizational ability and evident photographic memory served him well.

Binomial nomenclature. At and before Linnaeus' time, the name of the plant, as stated above, was the polynomial description, i.e., "*Trifolium* capitulis umbellaribus, leguminibus tetraspermis, caule repente." (This is white clover, from our lawns.) The chemists have never escaped this awkward procedure; for example, the name 2,4-dichlorophenoxy acetic acid (to pick a short example) is not only the name but the description of the compound. Probably you know that this substance is a weed killer that has the common name 2,4-D. But obviously, there is no such thing as a common name for most of the thousands of compounds that chemists and biologists write about; authors thus usually are forced either to use grossly elongated names or to invent acronyms. Sometimes writers independently come up with different acronyms.

The botanists, starting 200 years ago, have been smarter than the chemists. For Linnaeus placed a "trivial" name, a single word, in the margin of the page, this for white clover being "repens," the basis for the binomial *Trifolium repens*. (The reproduction of the subject Linnaean page shows this.) This idea caught like wildfire, and the description and the name (the binomial) became separate entities.

Descriptive language. Linnaeus developed the language of biological, especially botanical, Latin. Latin was, of course, the universal language for learned writing, but that of Cicero was scarcely applicable. Many new words and ways of saying things had to be devised. The language that Linnaeus polished was used in most taxonomic writing for 100 years after his death with some afterglow to the present; for example, a recent *Flora of Iran* is written entirely in Latin. And Latin is yet required for descriptions of new taxa.

Let's move now from this listing of the innovations and procedures of the scribbling Swede to some consequences for the present. First remember that Linnaeus' *Systema Naturae*, which went through 12 editions, 1735-1768, classified everything, animals and minerals as well as plants. Although the impact of his dicta here was less than that for plants, Volume 1, *Animalia* (1758) of the 10th edition of the *Systema* is the beginning date for nomenclature of animals. But I shall discuss only his cognate *Species Plantarum*, two editions, 1753 and 1762-1763.

Page from Linnaeus' *Species Plantarum* that includes *Trifolium repens*.

Linnaeus' *Species Plantarum,* first edition, 1753, is the beginning point for plant nomenclature for most plants. (The mosses and most of the algae and fungi leak in later with subsequent authors). What does this mean in the nomenclatural sense? In the decades after Linnaeus, the idea of priority (i.e., that the name first given to a plant should be the one used hereafter) was accepted in a general way, but not everybody agreed, and there were multiple interpretations of conditions affecting publication of a name. Anyway, nomenclature fell into such chaos that in desperation, an international conference in Paris was called in 1867 in which the initial "Laws of Botanical Nomenclature" were formulated and which designated Linnaeus as the beginning. Subsequently this proved too ambiguous, and ultimately the Code specified the first edition of *Species Plantarum,* 1753. This incidentally was a bad choice, because the second edition, 1762-1763, that corrected many mistakes and ambiguities would have been better, but we are stuck with the first. Then if Linnaeus' names are to be used, it is necessary to be sure what plant he was talking about. Maybe his brief description of *Trifolium repens* (white clover) was sufficient to distinguish it from any other *Trifolium* at his time, but because innumerable other *Trifolium* have been described since then, how can one be sure it was not one of them? Thus, his plant must be clearly identified (typified). But suppose, his description, specimen(s), and citations suggest different entities? These and other complications have produced generations of nomenclatural scholars, legalistic Linnaean lawyers, admiring, even adoring, analysts of the Master's methods and materials, akin almost to those generating the interminable literary criticism and adulation of Shakespeare. Regrettably they have not always agreed about some plants, of which the unfortunate soybean may be a prime example. Linnaeus originally had the genus *Glycine* in the eight members of his *Species Plantarum*, but the soybean was not among them. Rather two forms of it were described as *Phaseolus max* and *Dolichos soja*. How it eventually came to *Glycine max* (L.) Merrill is in itself a Ulyssesian odyssey.

Regarding much of this then as a by-product, how do we evaluate Linnaeus? There is no doubt, because of his ingenuity, his industriousness, and his effectiveness as a PR man, that he put taxonomy into big time, the dominant arena of biology for 100 years. But did he contribute any major new ideas as for example, those enunciated by **Mendel,** Darwin, or Pasteur? Except in methodology, I think not. He emphatically is not the father of modern classification as many presently assert. His artificial system was but a temporary blind alley—the flow chain of natural classification being from his predecessors, Ray and Tournefort, to **de Jussieu** in 1789. Linnaeus started at the bottom (the species) and worked up. By contrast his contemporary, the expansive Comte de Buffon (1707-1788) started at the top and worked down. He saw the grand scope of life, had a groping concept of evolution (subsequently censored), and was contemptuous of the Linnaean accounting approach. And so were some others. It is said that Linneaus recognized that his was an artificial system, and that he had notions of a natural classification to follow. Perhaps so, but life was not long enough.

However this may be, there are few who directly or indirectly have ever had more impact on botany and biology than Linnaeus. So *finis*—up and over.

SELECTED BIBLIOGRAPHY

Lindroth, S. 1973. C. Linnaeus. In: Dictionary of Scientific Biography. Vol. 8, pp. 374-381.
Morton, A.G. 1981. Linnaeus. In: History of Botanical Science, pp. 259-276, 281-283.
Stafleu, F.A. 1971. Linnaeus and the Linnaeans. Oosthoek, Utrecht, Netherlands. 386 pp.
Stearn, W.T. 1955. An Introduction to Species Plantarum and Cognate Botanical works of Carl Linnaeus. In: Facsimile of the First Edition of Species Plantarum, 1753, pp. xi–xiv, 1-176. The Ray Society, London.

Victor Albrecht von Haller (1708–1777)

Haller was a tempestuous, many-faceted genius. His principal research was in animal physiology and anatomy. But he also wrote on botany and natural history; he produced fiction and poetry; he participated in politics and public service. Even though botany was only a small part of his life, Haller left sufficient footprints in the igneous conglomerate of time that we need to stop and look at them.

Haller, Swiss, achieved a medical degree at age eighteen. He was then also climbing the Alps, collecting plants, and observing their habitats and variability. A succession of teaching, industrial, government, and political positions, mostly in Bern, represented the facade of his life there on.

Married three times, Haller's alliances produced (numerous?) children of whom eight lived to maturity. He claimed poor health, and the formidable list of ailments attributed to him suggests that he should have died by age twenty; instead he lived to sixty-nine, becoming enormously fat in his latter years. He was irritable, short-tempered, petty, and often antisocial (how could he have been other than antisocial considering the bulk of his writing?). His irascibility is evident in his publication and correspondence; he was involved in countless controversies because he could not tolerate the fumblings of lesser intellects. He vigorously participated in the then current heated squabbles about epigenesis vs. preformation (the embryo develops from undifferentiated tissue vs. the embryo in miniature is already there), and ovism vs. animalculism (the egg contains the inherited material vs. it is the sperm that brings the next generation). These were of the times as are biotechnology, molecular, and cellular biology presently.

A Swiss flora (*Historia Stirpium Indigenarum Helvetgiae*) was perhaps Haller's most important botanical publication. The ideas on which it was based were probably more significant than the flora itself. **Linnaeus** and Haller were almost exact contemporaries. Linnaeus believed in the fixity of species. Haller, on the basis of his continuing fieldwork, emphatically did not and was constantly throwing rocks at the crown of the emperor of botany, Linnaeus. Haller's ecological and experimental criteria for determining the taxonomic status of variable plant groups lacked only cytogenetics to resemble what we today call biosystematics. Thus Haller was so far ahead of his time and out of tune with his contemporaries that deserved recognition, at least in botany and in our posterior view, has fallen short.

Keep in mind, furthermore, while everyone else was taken up with the beauty and convenience of the Linnaean sexual system and his binary names, Haller would have none of it. He continued to use the pre-Linnaean diagnostic phrases to double as the names of the species; he thought the Linnaean classification artificial, and he did not like Linnaeus' genera. This was sad because his Swiss flora was beautiful. It described the country, climate, habitats, and ecology, was illustrated, and described the plants. But it was crucified by its ponderous and dated nomenclature and its dissentient classification. Although the supporting Haller philosophy was prophetic, it was passed by. On the other hand, looking to the past rather than the future, Haller's *Bibliotheca botanica* (1771-1772) drew continuing praise. It was an erudite, annotated chronological listing of all (so it is said) botanical publications to his time, and was much used.

Of Haller and Linnaeus, Stafleu (1971) has well said, "The two shared an inclination not to underestimate their own importance." The difference comes in that while Linnaeus was hailed almost as a savior, Haller lost out.

SELECTED BIBLIOGRAPHY

Hintzsche, E. 1972. Haller. In: Dictionary of Scientific Biography. Vol. 6, pp. 61-67, with citations.
Stafleu, F.A. 1971. Albrecht von Haller. In: Linnaeus and the Linnaeans, pp. 244-250. Oosthoek's Uitgeversmaatschappi, Utrecht, Netherlands.
Stafleu, F.A. and R.S. Cowan. 1979. Haller. In: Taxonomic Literature. Vol. 2, pp. 24-29. With citations; this is the best compedium for botanical references to Haller.

Michel Adanson (1727–1806)

Adanson proposed the first natural (by current definition) classification of flowering plants. His voice, contemporaneous with that of the mighty Swede, **Linnaeus,** was but a cry in the wilderness and was swept into oblivion by the overwhelming popularity of the Linnaean system and its simplified nomenclature. Now, after 200 years of floating in the waves, Adanson has been washed upon the shores and grasped by those who proclaim themselves as neo-Adansonians.

Adanson, French, was a loner, possibly a cantankerous one. He had the usual classical training, essentially free of natural science. But biology quickly became his big thing, and as a young man he spent some years working in Senegal, tropical Africa. On returning, as **Darwin** from the voyage of the Beagle, he came laden with specimens of all kinds, a journal, and ideas already running over into manuscripts, including a revision of the Sengalese mollusks. (Throughout his life, Adanson was into many things besides plants.) But most importantly to us, the Linnaean system didn't work in the tropics where he had been.

I have seen essentially nothing about Adanson's personal life beyond the fact that he apparently never tried to make a living in any conventional way and consequently experienced periodic periods of financial crisis. In his later years, he was awarded a small pension. Apparently he remained a bachelor.

But back to Adanson, a disembodied spirit. His polemic, *Familles des plantes,* two volumes, was published in 1763 and 1764. Volume 1 was an exposition of his principles and Volume 2 their application in a classification. He challenged the Linnaean or any classification based primarily on the flowers, and he debunked the Linnaean essentialism that it is the character that makes the group. Rather he espoused the thesis that one must consider all features of the plant and its biology: "There is no doubt that there can be only one natural method in botany, and it is that one that considers all of the parts, qualities, properties, and faculties" of plants. Adanson's "natural system" then is based on tabulation of numerous characters and crude (to us) correlation. Whether he used weighted or unweighted characters has been a topic of contention (Jacobs 1966). Sokal and Sneath, the primary neo-Adansonians, vigorously espouse unweighted characters in presenting their numerical methodology, "phenetic" taxonomy, which they carefully distinguish from "phylogenetic" taxonomy, the latter viewed as speculative spinach. Be this as it may, I think Adanson would have approved of my taxonomic aphorism, "It is not what a character is, it is how it behaves."

In short, Adanson was directly addressing one of the several major philosophies of modern evolutionary taxonomy. An important difference between Adanson and present brave Turks is that they are able to consider many more characters than were available to him, and they can investigate character relationship through the kind of numerical methodology unheard of until the last twenty-five years.

Adanson did not stop with sweet theory; he put his system to work in the second volume of his *Familles,* which included 58 plant families, 52 of them being flowering plants. This was the first treatment in which an explicitly defined natural classification was used in producing an interpretation of the plant world, crude it is true, but well in advance of any of its pre-Linnaean ancestors.

And this was not all. Adanson questioned the stability of species. He is said to have been the first to use the word mutation. Does a natural classification of multiple affirmatives express a genetic relationship? Adanson was essentially the first of the classification philosophers who were raising these questions (usually cautiously) in the century-long nascent ferment preceding Darwin.

But Adanson was ignored primarily because he refused to take up the convenience of either the Linnaean system or its nomenclature. He achieved (as we see it in posterity) botanical immortality primarily through the acceptance of his principles by **A.L. de Jussieu,** the merits of whose Natural Classification (*Genera Plantarum,* 1789) were such that it rapidly soared to ascendency over the Linnaean essentialism.

While the publication of the *Familles* may have been Adanson's high-water mark, he by no means rested on his laurels. His most massive publication was some 400 articles contributed to the supplement of Diderot's *Encyclopédie.* These then stopped abruptly but manuscript preparation continued. It has been presumed that he intended to produce his own encyclopedia, but this never came about.

And Adanson. He then lived long enough to be forgotten, except in *Adansonia,* botanical journal, and *Adansonia* L., the Baobab tree genus. One reason for his two-century obscurity is the fact that his papers (manuscripts, herbarium, and library) remained in the hands of the family for this extended period of time, becoming available to society, primarily the Hunt Botanical Library, only about 1960 (Stafleu 1971). Informal family care for this kind of material for the required number of generations must set some kind of a record. It also suggests the probability that Adanson was married and had children, but I have seen no mention of such among the several references cited here. The Hunt Institute then went into high gear (Lawrence 1963-1964) in his reincarnation. Take note, those of you who feel unappreciated. Maybe 200 years from now you will be recognized.

SELECTED BIBLIOGRAPHY

Jacobs, M. 1966. Adanson—the First neo-Adansonian? Taxon 15:51-55.
Lawrence, G.H.M. ed. Adanson. 1963-1964. 2 vols. Hunt Monograph Series. J. Cramer. Facsimile reproduction of Adanson's *Familles des plantes* with current commentary and annotation, especially by Stafleu. Not seen; a description of that publication is provided by Jacobs (1966).
Nicolas, J.P. 1981. Michel Adanson. In: Dictionary of Scientific Biography. Vol. 1, pp. 58-59, with citations.

Sneath, P.H.A. and R.R. Sokal. 1973. Principles of Numerical Taxonomy, pp. 23-24. W.H. Freeman.
Sokal, R.R. and P.H.A. Sneath. 1963. Principles of Numerical Taxonomy, pp. 16-18, W.H. Freeman.
Stafleu, F.A. 1971. Michel Adanson. In: Linnaeus and the Linnaeans, pp. 310-320. Oosthoek Utrecht, Netherlands.
Stafleu, F.A. and R.S. Cowan. 1976. Michel Adanson. In: Taxonomic Literature. Vol. 1, pp. 9-11, primarily citations.

Johann Hedwig (1730-1799)

Hedwig was born in Romania but spent most of his life in the German state of Saxony. A bryologist, he was the first to successfully elucidate sexual reproduction in the cryptogams. We are told that Hedwig was fascinated by mosses while yet a child. Since a study of human ailments was more likely to produce a living, he studied for the conventional M.D. at the University of Leipzig. Lacking money, he was grubstaked for some three years by a botany professor. He received the medical degree in 1759. Then he practiced medicine for about twenty years, however keeping up his botanical avocation by going out to botanize for several hours in the morning before visiting his suffering patients, and by studying his collections in the evening (Richards 1972). He received help in becoming a botanist by accumulating a small library through the kindness of the botanist Schreber. He was also given a microscope. Although of no moment, the microscope matter is of interest. Stafleu (1971) says that Schreber gave him the microscope as well as books. Stafleu adds that the microscope was a klunker magnifying no more than 50x, and that Hedwig gradually improved it, eventually achieving nearly 300x. But Richards (1972) says the microscope came from an inspector of instruments, J.G. Köhler. Maybe someone can get a master's degree by looking into the matter. Anyway, Hedwig was now fitted for business as an amateur.

During this period, as often conventional, he married. The union rapidly produced six children. His wife's death and subsequent family maintenance competed with the mosses for a few years. Richards (1972) says friends advised him to marry again. So he did, and six more children came into the world—sadly they all died before reaching maturity.

Hedwig's publications ultimately attracted sufficient attention that he acquired, by a succession of steps, a professorship at the University of Leipzig, including direction of the botanical garden. Richards (1972) has said after "years of poverty and neglect" honor came to him. I don't know about the poverty, but it strikes me that he was not neglected; rather he had more helping hands along the way (including those of his second wife) than most. Whatsoever, he now graduated to international recognition in the conventional way, i.e., election to diverse scientific academies and societies in several countries.

So what did Hedwig do? I repeat, he was the first to firmly establish sexual reproduction in the Cryptogams. A careful and skilled microscopist and biological artist, he replaced prior speculation with evidence. He discovered moss antheridia and archegonia and saw male gametes. He observed the germination of spores to produce the protonema (the filamentous initial growth of mosses). He completed the bryophyte life cycle. His *Species Muscorum Frondosorum* (1801), published shortly after his death, was an encyclopedia of nearly all mosses then known. That work presently serves as the beginning point for nomenclature of mosses except for the sphagnum group.

Hedwig also worked, but less successfully, with miscellaneous fungi, algae, and ferns. The ferns baffled him because he reasonably took the sporophyte plant to be the equivalent of the evident plant in the mosses, and, of course, failed to find any sex organs on it. Another fifty years were required before Hofmeister straightened this out. Continuing to look for sexual reproduction in the Cryptogams, Hedwig could make little of the fungi, but among algae he identified the antheridia and oogonia in *Chara* and illustrated conjugation (fusion of identical sex cells) in *Spirogyra*. He proposed and reasonably established the notion that sexual reproduction is widespread in the lower plants as well as among the seed

Johann Hedwig

kinds. Other botanists then were soon hunting for gamete (sex cells) production in these plants, and piece by piece, life cycles began to fall into place.

Hedwigia de Beauvois is a genus of mosses and *Hedwigia* is a journal for Cryptogams.

I have seen nothing to challenge the expressed thesis that Hedwig was a good man whose work was directed toward better comprehending the ways and wisdom of the Creator.

SELECTED BIBLIOGRAPHY

Morton, A.G. 1981. Hedwig. In: History of Botanical Science, pp. 322-324, 356-357. Academic Press, London.
Richards, P.W. 1972. Hedwig. In: Dictionary of Scientific Biography. Vol. 6, pp. 218-220.
Stafleu, F.A. 1971. Hedwig. In: Linneaus and the Linnaeans, pp. 254-255. Oosthoek's Uitgeversmaatschappi, Utrecht, Netherlands.

Jan Ingenhousz (or Ingen-housz) (1730–1799)

The approach to an understanding of the relationship between photosynthesis and respiration, initially hampered by the phlogiston paradigm, came stepwise and through the work of different hands and minds.

Priestley had set the stage in showing that plants could purify phlogisticated air (i.e., they produced oxygen), but attempts to duplicate his work met with inconsistent results. Perhaps he would subsequently have found the uncontrolled variable, but Ingenhousz beat him to it.

Ingenhousz was a Dutch physician who monkeyed with plants and electricity. Born in Brenda, The Netherlands, he received his M.D. in 1753. After further studying medicine and anatomy plus some physics for several years, he returned to his birthplace and practiced medicine. Then he went to England where he met Priestly and Ben Franklin (yes, the U.S. Ben Franklin; the two became friends), and Franklin interested him in electricity. Professionally, he learned a new inoculation procedure for smallpox, differing from that of Jenner's vaccination. Smallpox was then a common fatal disease. His success was such that he established an international reputation and was called to the royal court of Austria to provide his skill for its numerous members. He successfully did so and became personal physician to the Empress Maria Theresa. And

Jan Ingenhousz

there he stayed for some twenty years, his botanical contacts being limited to the fact that he married into a botanical family, daughter of the Austrian botanist N.J. Jacquin and sister of J.F. Jacquin.

Perhaps Ingenhousz wanted to get away from Maria Theresa, or from smallpox, or perhaps papa-in-law interested him in something different. For whatever reason, he took a vacation in England in 1778 and conducted numerous experiments on oxygen production by plants. And he published his findings the next year in a book with a title as long as the list of authors in some current papers in the journal *Science*. We will abbreviate the title to *Experiments upon Vegetables*.

Ingenhousz stood directly upon Priestley's shoulders in several ways. Among them he used the oxygen-measuring apparatus designed by Priestley to which he gave the name Eudiometer, praising it as "an instrument by which we may judge the degree of salubrity of the common air." The procedure was simply the injection of "nitrous air" (nitric oxide, NO) into a closed container over water. This compound, scarcely soluble in water, would react with atmospheric oxygen, producing NO_2, which is soluble in water; the rising of water in the glass chamber could then quantitatively measure the amount of oxygen that had been there.

And what Ingenhousz forthrightly demonstrated, again and again, was that light was needed were the plant to purify the air. Not just any part of the plant would do the job, neither the roots, flowers, nor the fruits, rather only the green leaves. Not only would this not occur at night, but the purification of the air would be greatly reduced in dim light and would rapidly cease at the end of the day. He tested his results on numerous kinds of plants under varied conditions, invariably finding, in his words, "All plants possess a power of correcting, in a few hours, foul air, unfit for respiration; but only in clear daylight, or in the sunshine."

Derivative from this work, he subsequently challenged the then current thesis of the French chemist, Hassenfratz, that plants get their carbon from the soil. Ingenhousz said his work demonstrated that it came from the CO_2 in the air. Also, as a physician, he wondered if patients with respiratory ailments could be helped by the use of oxygen. He developed equipment to provide it but apparently stopped there. Others picked up with the idea of medical use of oxygen. As for miscellaneous otherwise, he followed up on ideas from Franklin, doing pioneer work on the nature of electricity—but that is for the physicists to write about. And according to Van der Pas (1973), while studying algae (Priestley's "green matter"), Ingenhousz first used coverslips for microscopic observation of liquid mounts and beat **Brown** to the discovery of Brownian movement. (See R. Brown if you don't know what it is.)

Returning now to photosynthesis: only green plants and only in light. This was his discovery, a major one indeed to which he limited himself with a minimum of further work or speculation. As he picked up from Priestley, others, most immediately **de Saussure,** took up the baton for the next mile down the track.

Ingenhousz led a full and successful existence. His excellent income from Maria Theresa was for life. Personally he seems somewhat the antithesis of his contemporary in botanical searching, Priestley, who was always challenging any and everything, and then having to run from the authorities. Ingenhousz worked with the system, made friends, and avoided trouble. But as with too many of us, with declining health and inability to travel back to his adopted country Austria, his latter ten years are said to have been unhappy. He died in England.

SELECTED BIBLIOGRAPHY

Harvey, R.B. and A.M.W. Harvey. 1930. Jan Ingen-housz. Plant Physiology 5:283-287.
Morton, A.G. 1981. Ingenhousz. In: History of Botanical Science, pp. 332-334. Academic Press, London.
Reed, H.S. 1949. Jan Ingenhousz, Plant Physiologist. Chronica Botanica 11:285-396.
Van der Pas, P.W. 1973. J. Ingenhousz. In: Dictionary of Scientific Biography. Vol. 7, pp. 11-16.

Joseph Priestley (1733–1804)

Plants purify air! Progress in understanding the processes of respiration and of photosynthesis required prior knowledge of oxygen and carbon dioxide as independent components of the air. Priestley was involved in the stories of both.

But first a word about his frantic existence. And "frantic" started early because he was first handed to his grandparents and then largely raised by an aunt. His education was mostly that of the traditional humanities, logic, languages, etc., plus training for the ministry. On the side he read on everything, and his first publications were on languages and oratory. For a while he was a teacher and indeed was subsequently awarded an LL.D. for his work in education. In science, he conducted extensive experiments on gases (where his botany leaks in), electricity, and optics. He wrote voluminously about the results, for example, of gases, some six volumes. He was a minister in several churches, but his heterodox views turned once-faithful parishioners away. He evolved towards Socinian principles that were one of the diffuse ancestors of Unitarianism. His excessively liberal views, expressed also in writing, soon got him in trouble with government authorities who regarded him as a dangerous radical (nonconformity was then a crime in England). In addition he was all for the American colonies during their revolt against mother England. His continued existence in England became so touchy that he and his family fled to the United States where he became a correspondent friend of Thomas Jefferson. He remained in the United States the rest of his days.

Priestley married in 1762; there were three sons and one daughter (maybe more). His wife Mary Priestley, although said to have delicate health, must have possessed inner strength. She fully supported his diverse career activities that, owing to constant moving, must have kept the family in continuous tumult.

But the reason for writing about Priestley here is his study of gases and of plants. He got into this before there was any real chemistry. Stahl, German, about 1700, in an explanation of combustion had said that a substance like wood contained phlogiston. When it burned, the air became phlogisticated. Black, Scottish, had obtained carbon dioxide by heating a calcium carbonate. The gas could then be fixed into a solid once more by exposing the residue, calcium oxide, to the air. Hence he concluded that carbon dioxide (or by any other name) was a normal component of the air, and he appropriately called it fixed air. **Helmont** (a part-time botanist), more than a century earlier, first suggested that air is not just a simple entity, but rather a mixture, one of the components being spirit of the wood. **Hales,** the first plant physiologist (in 1727) suspected that plants "do imbibe elastick air," but he had no evidence. All of these investigators were talking about CO_2 under a variety of names. Their views reiterated that air was not a simple entity but rather a mixture in which the CO_2 was one of the components.

Priestley went from there. Partly as a result of living next to a brewery he became familiar with CO_2. Then he discovered that the heating of mercuric oxide produced, using Priestley's Stahlian terms, dephlogisticated air (oxygen). It made mice happy, would make a candle burn more brightly. If he exposed the mouse or a candle in a closed container to the phlogisticated air, the former would die and the latter would go out. Then if he put plants in his containers, he found they would change the phlogisticated air back to the dephlogisticated air, and again a candle could burn and a mouse could breathe.

Priestley's observations provided a starting place for others. And they came. Priestley had shared his goodies with Lavoiser, who ran off with them, naming the gas Priestley had discovered, oxygen, and defining the chemical changes in combustion and its biological equivalent, respiration. (Priestley never accepted Lavoiser's chemistry.) And **Ingenhousz,** the Dutch physician and botanist, confirmed Priestley's findings but added the fundamental qualification that CO_2 absorption occurs only in the light.

Morton (1981) has stated that Priestley's "researches on the nature of gases . . . were the experimental foundation on which modern chemistry was built." Well maybe, but I think Lavoisier is a better candidate for the first chemist. While you are thinking this over, I will add that Priestley was the one who gave the name *rubber* to gooey stuff imported from the New World because it would rub out pencil marks.

SELECTED BIBLIOGRAPHY

Gibbs, F.W. 1965. Joseph Priestley. Nelson and Sons, Camden, N.J. 258 pp.
Morton, A.G. 1981. Priestley. In: History of Botanical Science, pp. 329-332. Academic Press, London.
Schofield, R.E. 1975. J. Priestley. In: Dictionary of Scientific Biography. Vol. 11, pp. 139-147.

Sir Joseph Banks (1743–1820)

One cannot read, even casually, of plant science in the latter eighteenth century without repeatedly encountering the name of Joseph Banks. But Banks published essentially nothing, and by conventional reckoning, was a nobody. Rather, he was a combination of a one-man National Science Foundation and president of the American Institute of Biological Sciences for forty-two years. He was an organizer, patron, entrepreneur, manager, friend, and fiscal supporter of botanists, unmatched by any other in his days or perhaps in all time.

As a boy, Banks became enamored of the beauty and the ways of nature; he was interested in anything "that moved, or crept, or grew, or stood still" (Rogers 1988, p. 603). He went to Oxford, but the professor of botany, Sibthorp, never gave lectures, so Banks quit. Then he inherited the family fortune. And it was his money, combined with his enthusiasm, activism, and ability to make friends in high places that were the magic potion. Bank's professional disposition is reasonably portrayed in his accomplishments; he was an operator cum laude. Some slight notice of his personal life may be found because he is the subject of several biographies. He is alleged to have had a mistress, Sarah Wells, in his younger days. At age thirty-six he married Dorothea (first name is all I have) of whom one hears no more except that there were no children. Apparently his

Sir Joseph Banks

sister Sophia was a part of his household, and (by implication) there dominant (Rogers 1988). Beyond his botanical management, Banks had his own sixty-two farms of which to take care. His home avocations there were sheep and cattle breeding.

At the age of about twenty-five, Banks wangled the position of naturalist on the Captain Cook Endeavor expedition to the South Seas. It was certainly the supremely scientific trip of its time. For Banks took along some eight assistants and extensive collecting equipment for obtaining, catching, and preserving everything from ocean life to grasshoppers and palm trees. He paid the bill for the whole business, estimated at 10,000 pounds. The collecting and journal recording of everything (botanical, geological, and zoological) that Banks did probably represents his only direct botanical accomplishment. But his journal was not effectively published until 1962.

Banks became president of the Royal Society of London in 1778 at age thirty-five and so remained until his death in 1820. In this role, it has been said that he dominated the scientific community as Samuel Johnson did literary circles. Soho Square, Banks' home, was a meeting place for scientists from the entire European continent, and Banks' unexcelled library and personal herbarium (the largest of any in Great Britain) were available for reference. He funded the activities and daily bread of various participating botanists. He was one of the founders of the Linnaean Society of London. He maintained contact with continental botanists even when England was at war with their countries. The voyage of the Bounty (*Mutiny on the Bounty* of current knowledge) was another Banks operation. The mission was to bring breadfruit from the eastern tropics for trial in the West Indies.

Banks also wanted a botanical garden. Kew Gardens, a few acres, was then apparently a pretty flower collection for the Royal Family. Banks was a friend of the king, George III. George appointed Banks the manager. Banks used the garden as a living plant research resource for hundreds of kinds of plants from all over the world. This was the genesis of the Kew that presently exists.

Botanical literature contains some five genera in five families bearing the name *Banksia*, (Zingiberaceae, Thymeliaceae, Rosaceae, Lythraceae, and Proteraceae). The last, an Australian group of shrubs and trees (a conserved name) is the one presently used.

The ultimate custodian of Banks' affairs was **Robert Brown,** one of the major British botanists of the time. Everything was willed to him upon Banks' death with the proviso that it subsequently go to the then fledgling British Museum, which it did, initiating the establishment of a Department of Botany that yet exists.

The half century, 1770 to 1820, was the Banksian Era in British botany.

SELECTED BIBLIOGRAPHY

Carter, H.B. 1988. Sir Joseph Banks: 1743-1820. British Museum, London. 671 pp.
Foote, G.A. 1970. Joseph Banks. In: Dictionary of Scientific Biography. Vol. 1, pp. 433-437, with citations.
Rehbock, P.F. 1981. Science 212:319-320. Review of: C. Lyte 1980. Sir Joseph Banks, 18th Century Explorer, Botanist and Entrepreneur. David and Charles, North Pomfret, Vt. 248 pp.
Rogers, P. 1988. Sir Joseph Banks. Times (London) Literary Supplement, June 3-9, 1988, pp. 603-604.

Jean Baptiste Pierre Antoine de Monet de Lamarck (1744–1829)

Lamarck might be remembered for his polynomial name, but it is usually shortened to Jean Baptiste de Lamarck. Otherwise he is usually recalled for the wrong reasons, for example, he was the chap who said that the giraffe's neck became longer because said animal was constantly having to stretch it to get yet higher leaves. In a different realm, botanists who read authorship of plant names frequently find "Lamarck" or abbreviated to "Lam." He is known to them even if they have not thought about giraffes. Lamarck's initial writing, it is true, was of systematic botany, and he described numerous new species. But he rapidly became a visionary biologist who sought the evolutionary mechanisms of the living world. And now, Stafleu's short biography (1971) is titled *Lamarck: The Birth of Biology*. That is why he should be better known.

Jean Baptiste, the youngest of eleven children came of a family of the French "impoverished nobility." His elder brothers, following his father, were trained for a military career. Perhaps because it was cheaper, the priesthood was selected for him, and he was sent to a Jesuit school. He didn't like it; he quit at age sixteen and joined the military on his own. During his army career, which lasted about six years, he was shipped around Europe and became interested in the diversity of vegetation of that continent. As a pastime, he collected plants and became an amateur botanist. Then for several years, he was a bank employee in Paris, simultaneously writing a three-volume *Flora Française* (1779), which proved (for its subject) to be a best-seller. It attracted sufficient attention that Lamarck obtained a position as botanist at the Jardin du Roi. When the French Revolution upended everything, botany per se lost out because Lamarck emerged as professor of zoology at the Museum d'Histoire Naturelle, the new name for the Jardin du Roi. Specifically, he was supposed to study "worms and insects." No matter, he rapidly became a natural philosopher in the broadest sense of the phrase, his writing including meteorology, geology, and chemistry, as well as evolution (but the latter not by that name).

Seemingly he was settled for life, which should have been one of those "and they lived happily ever after" affairs. It wasn't. The pay was poor and erratic, and he and his sequential families suffered poverty and successive catastrophes. He survived some three (possibly four) wives, the first of whom, so I read, he married on her death bed subsequent to fifteen years and six children. By the time of his latter years he was overtaken by the younger generations and rapidly forgotten. However, he continued to work despite deteriorating health and eyesight. Blindness was complete by 1818, and he dictated to one of his daughters who stayed with him. His death in 1829 was essentially unnoticed. Money for a funeral had to be raised by a poverty appeal. The yet single daughter who had cared for him was left with no support; one wonders what happened to her.

Lives are ephemeral and commonly painful. Let us turn instead to Lamarck as the biological activist and writer. Initially he was an eighteenth-century fundamentalist who accepted the Linnaean precept of fixity of species. He soon broke with **Linnaeus.** In the first volume (really a textbook) of his *Flora Française* (1779), he also disavowed the Linnaean essence, a single character being the mark of the group. Rather his playing cards were those of **Adanson** and **Jussieu,** which asserted that the natural affinity of taxa must be based on multiple characters. Beyond sweet theory, the success of the *Flora Française* was based on its lucidity: excellent descriptions, the innovation of dichotomous keys

Jean Baptiste de Monet de Lamarck

for identification, and the fact that it was written in French rather than Latin. It was followed by several botanical books among which three volumes of the tremendous *Encylopédie Methodique* of Panckoucke were perhaps most significant.

I repeat that Lamarck was a broad-scope natural philosopher. It has also been stated that he was uninterested in detail. The first indeed is the quintessence of Lamarck. The latter is not correct, at least for his botanical publications despite his quoted disparagement of "little details." There his descriptions of species and tabulation of their distribution and habitats were meticulous. His attention to detail was also evident in his discernment and naming of numerous new species. And I am convinced, although I cannot document it, that from his botanical work came the genesis of the thesis that natural affinity means genealogical relationship.

After the revolution when new organizations were substituted for the old, and when Lamarck was professor of the insects and worms, he promptly conjoined these creatures with other groups under the name of invertebrates. During this period, his work in chemistry and geology changed him from a vitalist to a mechanist and gave birth to his concept of a chain of life. In geology he was evidently a Hutton proselyte of uniformitarianism, perhaps even an unheralded Lyell (the so-called father of geology) ahead of his time. As a side issue, his activist meteorology pronouncements resulted in the establishment of weather recording stations across France, but these were subsequently disapproved by Napoleon and closed down.

The most evident primary thrust of the latter part of Lamarck's life culminated in his *Histoire naturelle des animaux sans vertèbres,* 1815-1822, seven volumes, which placed in reasonable order what before had been but chaos. It also emphasizes the idea that Lamarck never wrote anything you could carry about without a push cart.

It is more difficult to follow the pathways leading to Lamarck's premise of the mutability of species through time. Let it be said immediately that Lamarck was by no means the first person prior to **Darwin,** who had toyed with this disturbing thought. For example, Erasmus Darwin, grandfather of *the* Darwin, mused about the topic, and the expansive **de Buffon** had presented it in his multivolume *Natural History* in the middle of the eighteenth century (like Galileo, he had to retract). But Lamarck was the first to develop the evolutionary thesis as a documented, serious proposition.

All right, if there is a chain of life, what and how? Lamarck's basic premises, drawn from both his botanical and zoological background, seem to have been about as follows: (1) All life tends to change from the simple to the complex; (2) The direction of such changes and their nature as they affect homologous organs are mediated by the environment, which changes with time; and (3) Such changes in organs are subject to the use-disuse principal; those that are most favorable for survival (i.e., most used) become those inherited. In total they constituted a new conceptual paradigm.

Through selective magnifying mirrors, comes of course the presently ridiculed neck of the giraffe, which in Lamarck's writing was only a secondary example. And was it really ridiculous in his time? No. Many contemporary biologists already believed in the inheritance of acquired characters. Should it be regarded as ridiculous now? Again no. Both Lamarck and Darwin more than a century later, without the benefit of genetics, had to hypothesize concerning the nature and thrust of variation. Use-disuse was evident to Lamarck; Darwin partly bailed himself out with selection over time from the pool of natural variation.

Initially Lamarck's theses seemingly were received with interest and even some applause. True, his mechanism to propel change, "use results in inheritance," was caricatured by his detractors beyond his proposition but this was evidently no fatal wound. It took someone with propositions more in tune with the cultural fundamentalist climate of the time to derail Lamarck. That came from the brilliant and vigorous pioneer paleontologist and zoologist, Cuvier, who though accepting the antiquity of the earth and its life, flatly rejected evolution in any form. He explained the existence of fossils through the thesis of catastrophism, which postulated successive creations and subsequent worldwide catastrophes wiping out all prior life. Humans and present life were merely the subjects of the most recent creation. Cuvier completely overshadowed Lamarck and passed him to the dust bin.

So, all in all, was Lamarck an abject failure? No, he was not.

Jean Baptiste de Monet de Lamarck

His work portrayed and even epitomized, within the intellectual growth of a single individual, the changing currents of the life sciences between the eighteenth and nineteenth centuries. This was transfiguration of a static world of perfect creation towards a dynamic one subject to change and diversification over time. Lamarck, starting with the former, then evolved in a way that no other biologist had done and produced the first orderly, reasonably documented explicit statement that the expression "natural affinity" means phyletic association.

Though glory and acclaim came not to Lamarck, one can perhaps assert with Stafleu (1971) that he created biology as a coherent discipline, complete with the name.

SELECTED BIBLIOGRAPHY

Burkhardt, R.W. 1977. The Spirit of the System: Lamarck and Evolutionary Biology. Harvard University Press, Cambridge, Mass. 286 pp.
Burlingame, L.J. 1973. Lamarck. In: Dictionary of Scientific Biography. Vol. 9, pp. 584-594.
Linnoges, C. 1978. Lamarck in his milieu. Science 199:1427-1428.
    Review of Burkhardt, cited above, with independent commentary.
Stafleu, F.A. 1971. Lamarck: The birth of biology. Taxon 20:397-442.

Antoine-Laurent de Jussieu (1748–1836)

The Jussieus, French, constituted a botanical horde. Among those of three successive generations, some five were botanists. Of these, the impact of A-L. de Jussieu, on the history of botany was the most substantial. That is because he was the first (1789) to formally propose a documented natural classification of flowering plants that quickly superceded the convenient but artificial system of **Linnaeus.**

Jussieu obtained his medical degree in 1770 and soon became affiliated with the Jardin du Roi, the King's Garden. Keeping a low profile, he managed to continue his work across the French Revolution, emerging as director of the "Museum" (I suppose this to be the Museum d'Historie Naturelle, the new name for the Jardin du Roi). There he stayed until retirement in 1826.

As early as 1774 he published an arrangement of plants derived from that employed in the garden plantings of his uncle, Bernard de Jussieu, which were arranged somewhat according to plant families as we now understand them (i.e., the mints together, the umbellifers together, etc.).

To document and express the ideas suggested in the garden, A-L. Jussieu drew on three sources: (1) Bernard Jussieu's thinking and the garden itself; (2) the vital inheritance from **Adanson's** *Familles* (1763-1764); and (3) herbarium holdings, not only those on the shelves of the museum but from anywhere else he could get

Antoine-Laurent de Jussieu

them.

Then (1789) he published his epochal *Genera Plantarum*. There he provided a synthesis of his methodology, most importantly a review of the Adansonian dicta that multiple characters (the genie, as he called them) are required to define a group. Thus he rejected the Linnaean essence thesis, namely that a taxon is identified by a single character. On the other hand he, most wisely, promoted the Linnaean system of binomial nomenclature that his predecessor, Adanson, had eschewed and that had crippled the usefulness and acceptance of his earlier system. Thus, *Genera Plantarum*, although deriving much from Adanson, succeeded and was hailed, whereas the more fundamentalistic or rigid Adanson's *Familles* had failed. And now in posterity, it is with the *Genera* (1789) to which a natural classification of flowering plant families is usually dated.

Jussieu's approximate contemporary, **Lamarck** picked up the Jussieu system intact in the introductory discourse to his volumes for the *Encylopédie Methodique*. Subsequently the generic frame, allowing the addition of more families and genera, has continued to the present. Seventy-six of Jussieu's family names are now conserved in the International Code of Botanical Nomenclature; eleven names come from Linnaeus; forty-six from his uncle Bernard, six from Adanson, and he, Antoine-Laurent, proposed some thirty-four names. This counting is taken from from Morton (1981), which seems in turn to be derived from Stafleu (1971).

So while Linnaeus provided binomial nomenclature, the name Jussieu is linked to that of a natural classification of plants. But always there must be some questions and qualifications. Were the constituents of Jussieu's character assortments weighted or unweighted? Inspection suggests that there was considerable weighting in his interpretations, i.e., he let his intuition guide uncertain steps. Happily he had the gift of good intuition at the level of genera and families. But they evidently failed him for the higher groups of more abstract nature (e.g., classes). Another qualification must be that the system dealt primarily with flowering plants; at his time the cryptogams could be but scantily fleshed.

Finally now one must remember Jussieu's idea ancestor, Adanson, who for 200 years hence, was lost in obscurity and only recently brought to light. Perhaps Adanson's *Familles* and Jussieu's

Genera should be considered two parts of the same.

SELECTED BIBLIOGRAPHY

Morton, A.G. 1981. A-L. de Jussieu. In: History of Botanical Science, pp. 311-313. Academic Press, London.
Stafleu, F.A. 1971. Antoine-Laurent de Jussieu. In: Linnaeus and the Linnaeans, pp. 321-332. Oosthoek, Utrecht, Netherlands.
Stafleu, F.A. 1973. A-L. de Jussieu. In: Dictionary of Scientific Biography. Vol. 7, pp. 198-199.

Thomas Andrew Knight (1759–1838)

Knight was a pioneer plant physiologist/horticulturist, many of whose diverse experimental activities were well ahead of his time; indeed some years passed before others carried further his banners of discovery.

Of a wealthy English family, T.A. Knight, after college, took over management, then ownership, of a 10,000-acre farm. Here he conducted both practical farm management experiments as well as those of pure inquiry. He reported much of the latter to **Sir Joseph Banks** who "read" them to the Royal Society of London and subsequently arranged their publication in the *Philosophical Transactions*.

Knight married; there were four children. Although he is said to have been a shy man who lived a retired life, he was president of the Horticultural Society of London for twenty-eight years. Possessed of a photographic memory, he could, upon appropriate provocation, quote page after page of classics such as Virgil and Milton. A portrait shows him to be clean shaven and bald headed.

Mr. Knight did not just manage; he tried to improve, and much of his work deals with practical horticulture. But his curiosity led him to rainbows beyond, and I give examples of two of these.

Why do roots grow down and shoots upward? Is it innate in plants or caused by gravity? If the latter, motion should effect or suspend it. So Knight rigged up an ingenious wheel to which imbibed seeds were fastened about the margin. Placed in vertical position, water power spun it approximately 150 rpm. The roots grew outward; the shoots to the middle, but after passing the center then curved back. Seemingly centrifugal force substituted for gravity. Then the investigator constructed a horizontal wheel capable of running up to 250 rpm. Radicles grew outwards but turned approximately 10 degrees down. The shoots bent the opposite direction. If the wheel were rotated at slower speeds, the declination from the plane of spin was correspondingly larger. Evidently the causes of the direction of growth are external forces, which in nature can only be gravity. Although most of this was new, i.e., the first real study of geotropism, Knight's wording seems to suggest that he knew he had only partly solved the problem: How does gravity make the plant so behave?

And the ever inquisitive Mr. Knight worked with the movement of sap. He assumed it to be of circulatory nature. This prevalent notion possibly derived from Harvey's conclusions concerning humans. He traced upward movement with a colored solution obtained from grape skins. The plumbing, he said, is in the alburnum (xylem). He knew of capillarity and transpiration but not of root pressure. He refused to speculate, saying that the movement of water to the leaves is generated "by what means I shall not attempt to decide." This was probably wise. His data and the knowledge of physics at his time scarcely allowed him to come to any firm conclusion. On the other hand, he ascribed the movement of the descending sap to gravity. It differs from that ascending "as a result of nutrients received from the leaves."

Although Knight is sometimes called a forerunner of **Mendel,** his work on peas seems a little less than glorious. He wanted to develop new apple varieties by crossing them. But he first turned to peas because they are annuals, thinking he might get some ideas that could then be applied to the slower generation crop. He observed dominance in seed color (grey over white). He backcrossed F_1 hybrids to the recessive and obtained both types again. Probably he crossed some of the hybrids with one another. But evidently not numerically inclined, he obtained none of Mendel's data. I doubt that his unsubstantive results helped his apple improvement; apples are highly heterozygous and the seed progeny rarely

Thomas Knight

resemble the parents. Anyway, it is a non sequitur to call him a forerunner of Mendel.

Although several authors label Knight as a genius, I am inclined to suggest rather that he was an orderly, ingenious, and methodical cuss with a phenomenal memory. He is to be credited for preparing the stage for several kinds of future performers. For that he deserves our applause.

SELECTED BIBLIOGRAPHY

Anonymous. 1913. Thomas Andrew Knight, 1759-1838. The American Breeders Magazine. 4:1-4.
Shull, C.A. and J.F. Stanfield. 1939. Thomas Andrew Knight. In memorium. Plant Physiology 14:1-8.
Simpkins, D.M. 1973. T. Knight. In: Dictionary of Scientific Biography. Vol. 7, pp. 408-410.

Christiaan Hendrik Persoon (1761–1836)

The dates of Persoon's birth and death cannot be precisely established (Stafleu and Cowan 1983). Persoon perhaps was **Linnaeus** of the fungi; he started the ball rolling, and it was then picked up by others. True, many conspicuous fungi, mushrooms for instance, had been described prior to Persoon, indeed back to antiquity. But the understanding of the fungi as a group prior to Persoon was nil. It is possibly epitomized by the fact that Linnaeus' aptly named genus *Chaos* included two species of fungi among six there described. And *Chaos* was listed among the *Vermes* (worms).

Persoon was evidently an odd duck, but since this is a common quality among botanists, it should not be held against him. He was born in South Africa; his mother died shortly thereafter. He was sent to Europe at the age of about thirteen with a small legacy to get his education. His father died within the year but arrangements allowed continuation of his educational support. Starting in theology, the young man transferred to medicine, and finally obtained not the usual M.D. but apparently an honorary Ph.D. (1799). Shortly afterwards (1802), Persoon moved to Paris where he lived the rest of his life on an upper floor of a house in a poor part of Paris. Seemingly he neither married nor had a salary-bringing appointment. It is not surprising then that he lived most of his life in poverty. He evidently was a recluse in the sense that he hid from people but was not a recluse in that he corresponded with botanists all over Europe.

Just how or why Persoon turned to plants is seemingly not known. One hazards that he found them more enjoyable than people. He was interested in the whole plant world and his *Synopsis Plantarum* (1805-1807) was much used because of its well-honed descriptions. The two volumes of this work are tiny as to page size (slightly larger than a three-by-five card) but thick and contained a summary of plants then known, approximately 20,000 species, in not quite microscopic print *("in parvo copia")* on tissue-paper-thin pages. Stafleu and Cowan (1983) say a simultaneous printing was on "heavy" paper (not seen).

But it was in the fungi that Persoon pioneered. And why fungi? Perhaps in style of life he felt a certain empathy with them. However as that may be, his continuing descriptions and classifications of these plants produced a succession of one to several volume works (Stafleu and Cowan 1983). His criteria for higher categories, based largely on macroscopic and hand lens features were, of course, arbitrary. But the classification presented a structure that could hold almost everything and be understood by others, as did Linnaeus' ordering of higher plants. It was at the generic level that both Linnaeus and Persoon excelled, and most of the genera they established are yet recognized today. Nomenclature for the Uredinales, Ustilaginales, and Gasteromycetes begins with Persoon's *Synopsis Fungorum* of 1801.

Persoon for many years corresponded with anyone interested in fungi, and he assembled the largest fungus herbarium then existent (and it still exists). Ramsbottom (1934) has given a sample of correspondence with J.E. Smith, a British botanist, that presents Persoon's fixation on obtaining specimens of anything mycological. Persoon's letters wander over three languages, Latin, English, and French. He recognized that his English, although easily read, was sometimes individualistic. He stated concerning a manuscript submitted for the *Transactions,* of the Linnaean Society: "You are allowed, in case You should meet with something contrary, or not convenient to the genius of English Idioms, to alter and mend it."

There is no doubt that Persoon was the first real mycologist.

After him came **E. Fries,** and after Fries was **de Bary,** and mycology was in full bloom.

SELECTED BIBLIOGRAPHY

Donk, M.A. 1972. C. Persoon. In: Dictionary of Scientific Biography. Vol. 10, pp. 530-532.
Ramsbottom, J. 1934. Proc. Linnaean Soc. London 146:10-21.
Stafleu, F.A. and R.S. Cowan. 1983. Persoon. Taxonomic Literature. Vol. 4, pp. 178-185. Extensive citation and publication list.

Johann Jacob Paul Moldenhawer
(1766–1827)

The description and interpretation of the structure of plants lay fallow for a century following the investigations of **Grew** and Malpighi. A whole century! Why so? I conjecture this was because the feverish excitement of botanists lay in the wonders and wealth of plants that exploration brought to their amazed eyes, and also because major improvements in microscopy did not come until the middle 1800s. But nevertheless several botanists now dabbled with plant structure. Except for Moldenhawer, however, their efforts primarily were but cosmetic decorations on the errors of the past.

Moldenhawer, German, son of a minister, studied theology and the classics. He eventually married and had one daughter. I have the impression he was a loner, but this is not explicitly documented. When and how he became interested in plant science is not evident. Of record, his first botanical publication was 1791, and in 1792 he turns up as "Extraordinary Professor of Botany and Fruit-tree Culture" of the faculty of philosophy at the University of Kiel. There among other things he taught Greek literature. From about 1795 until 1812, when his *Beitrage zur Anatomie der Pflanzen* was published, he devoted his extra (?) time to the study of plant anatomy. Then he quit and turned to fruit tree production.

Moldenhawer used a retting process to prepare his material,

i.e., plant parts were maintained in water long enough for the middle lamellae to disintegrate so cells could be separated through maceration. He introduced the equivalent of coverslips so that tissues could be examined in water. This allowed him, more clearly than prior workers, to identify the cellular units amid the conglomerate mess. He differentiated between two basic tissue types, vascular (Gefässbündel) and parenchymatous. He described and illustrated fibers and vessels—earlier workers had doubted that fibers were cells. He described vascular bundles and identified the cellular nature of the cambium noting that wood was internal to it and bast external. He correctly interpreted secondary thickening and annual rings. Stomata, he observed, were not cells with a hole in them, but rather openings or pores surrounded by two cells. He emphasized that every cell has its own wall independent of that of surrounding cells and assumed that there is a cement that holds them together—the middle lamella of subsequent terminology.

Cell theory by textbook dicta is credited to the **Schleiden**-Swann combination some years later. Perhaps this is reasonable, because Moldenhawer, primarily an observer, presented no explicit thesis that all cells come from pre-existing cells. But really neither did Schleiden; that decree awaited Virchow's *"omni cellula e cellula."* Cell theory or doctrine thus was scarcely a discovery by any one person; the evidence for it was a product of sequential work by several individuals, and its identification with any particular worker is arbitrary. I have the impression that Moldenhawer, rather than Schleiden, supplied the most important initial documentation, at least for plants, supporting such a theory.

After the production of his book, Moldenhawer abruptly turned to fruit tree culture for the rest of his life. Perhaps the authorities told him it was time to stop playing around and get busy with what he was hired to do. Too bad.

The small, Brazilian legume genus *Moldenhawera* Schrader is named for Moldenhawer. Fine, but seems a bit out of context.

SELECTED BIBLIOGRAPHY

Morton, A.G. 1981. J. Moldenhawer. In: History of Botanical Science, pp. 365, 368-370. Academic Press, London.

Wolf, J.H. 1974. Moldenhawer. In: Dictionary of Scientific Biography. Vol. 9, pp. 455-456.

Nicholas Theodore de Saussure
(1767–1845)

At the onset of the nineteenth century, Lavosier chemistry had opened up new opportunities for understanding the feeding of plants. Lavosier had defined the chemical constituents of carbon dioxide and established the nature of oxidation, thereby properly and finally interring the phlogiston flimflam. Also water had been synthesized from hydrogen and oxygen by Cavendish. De Saussure now became the first plant chemist.

De Saussure, French/Swiss, came of a well-to-do family. His father was a geologist and traveler, his grandfather a country gentleman. He was born in Geneva, Switzerland, and spent most of his life there except for a few years during the time of the French Revolution. On return, he accepted a position as honorary professor of mineralogy and geology at the Geneva Academy. But he neither published nor gave any courses on these subjects. Instead he occupied himself with plants. Oh yes, he married about 1796.

The middle and latter part of his life, de Saussure worked extensively on plant metabolism, particularly the chemical content of seeds and the changes during the process of germination, for example, the conversion of starch to sugar. He also investigated fermentation reactions. But in posterity, the most significant of his studies was the first, conducted while he was yet young. It concerned the nature of the gaseous exchange of plants with the atmosphere and was a major step beyond his predecessors Senebier and **Ingenhousz.** De Saussure's *Reserches Chimiques sur la Végétation* was published in 1804. It was gobbled up and went through several editions and translation into German. At the least, Saussure had everyone's attention.

Using careful quantitative methods, he showed that respiration in plants is equivalent to that in animals and that the amount of O_2 used is equal to that of CO_2 produced. With regard to photosynthesis, he corrected Senebier's conception that the CO_2 taken up is that dissolved in water. Working with known volumes of CO_2 in the air and green plants in sunlight, he found that the CO_2 loss from the air was equivalent to carbon fixation in the leaves. As a check against Senebier's conclusion, he also tried plant parts immersed in water. The carbon fixation matched the loss from the air, not from that dissolved in the water.

So, summarizing, the plant gives off CO_2 in respiration. Green plants in the light take in CO_2 and give off O_2 during carbon fixation. These are the interchanges with the air. All other nutrients come with water from the soil, including nitrogen, even though nitrogen makes up most of the air. With respect to mineral nutrition from the soil, de Saussure's experiments showed that the uptake of salts from the soil is not equivalent to their proportion in water. In short, he also first demonstrated that plants somehow absorb selectively.

See Morton (1981) for a descriptive analysis of Saussure's meticulous methodology, one that enabled him accurately to sort out the prior confusion attendant to gas exchange from the two opposed metabolic pathways, i.e., respiration and photosynthesis. It was keen work and nobody had to subsequently arise and default him.

Two plant genera were named for Saussure during his life time, *Saussurea* (a composite) by **de Candolle,** and *Saussuria* (a mint) by Moench. While this baggage seems scarcely to fit the man, it well suggests that he was known and honored.

SELECTED BIBLIOGRAPHY

Morton, A.G. 1981. de Saussure. In: History of Botanical Science, pp. 337-342. Academic Press, London.

Nicholas de Saussure

Pilet, P.E. 1975. de Saussure. In: Dictionary Scientific Biography. Vol. 12, pp. 123-124, with citations.

Saussure, N. de. 1804. Recherches Chimiques sur la Végétation (in part). English translation by M.L. Gabriel. In: Gabriel, M.L. and S. Fogel. 1955. Great Experiments in Biology, pp. 161-165. Prentice-Hall, Englewood Cliffs, N.J.

Robert Brown (1773–1858)

Most contemporary botanists have heard of Robert Brown. "Not me," you say. But no doubt you have encountered what is called *Brownian movement* in your physics. Brown, looking at pollen grains in water, noticed that they jiggled around irregularly. He tried nonliving particles of about the same size and found the same. He reported this in 1827, obviously without explanation, which had to await advances in kinetics.

In a way this epitomizes Brown. An unexcelled jack-of-all-botany in his time, he distanced all others in new observations and usually new ideas.

Brown, a Scot, was the son of a clergyman "of strong, independent views." He was medically educated and served five years as a surgeon's assistant in the army. The turning point in his life came when he won the esteem of **Sir Joseph Banks,** the supreme financial benefactor and action-promulagator of botany in his time. Banks arranged for Brown to go on the Flinders expedition, South Africa, Australia, and Tasmania, a five-year jaunt. What Brown saw opened up the world to him. He subsequently, probably supported by Banks, became a monographic-descriptive systematist. Then he went to a bit of everything else, writing on a variety of botanical topics, which I sample. He also acted as general caretaker for the Linnaean Society and in his latter days established a Department of

Robert Brown

Botany at the British Museum. Along the way he refused three professorships.

Brown never married. He was a loner in the sense that he stayed away from contention, students, and public affairs. But he liked the company of his peers, and **Asa Gray,** an American botanist, once said he was "very fond of gossip at his own fireside." We know more about his personality characters than many botanists because Brown kept a diary for many years and because an unusual number of his contemporaries wrote about him. He played the violin, he liked port, and he struggled with German verbs. **Darwin** said, "He seemed to me chiefly remarkable for the minuteness of his observations and their perfect accuracy." Martius, a German botanist, said that his imposing form was tall and slender. Asa Gray averred that Brown was "a singular looking man with a heavy lower lip and jaw, and generally carries his head down . . . he is a curious man in other things besides botany . . . in company he is silent and reserved . . . and he is the driest pump imaginable . . . one day at the Museum, Brown was there, but did little except read the newspaper and crack his jokes. Boott [secretary of the Linnaean Society] stated 'of all the persons I have known, I have never known his equal in kindliness of nature.'" These quotations are derived primarily from fuller exposition by Ramsbottom (1931). The various comments of Gray come from his letters.

Brown's collecting loot on returning from the Flinders expedition included several thousands of plant, animal, and mineral specimens. He set about working the plants up for a *Prodromus Florae Nova Hollandiae* (New Holland was then the name for Australia). He is said to have had about 140 new genera and 1700 new species and personally financed publication of Volume I of the *Prodromus.* It covered ferns, monocots, and some dicot families and sold abysmally. Brown finally took the remainder of the stock and gave the books away. Except for some miscellaneous publications, for example, revisions of the Proteraceae (South Africa and Australia primarily), the Asclepiadaceae, and miscellaneous this and that, it was the end of his systematic production. It may be presumed that the contemporary lack of interest in the *Prodromus* caused him to change the direction of his efforts. The matter puzzles me because, as at least one author says, he was the most outstanding British systematist of his day, and certainly he is highly regarded now.

Anyway there was plenty else for him to do; so much so, we can only sample his findings.

Brown studied the growth of the pollen tube to the micropyle, and the development of the ovule, thus tracing pollination/fertilization nearly to ultimate action (keep in mind that microtechnique was then primitive compared to the present). He was the first, I think, to provide a fundamental basis for distinguishing the Angiosperms and Gymnosperms—the difference in ovules and seeds. He asserted like Goethe that floral parts are modified leaves, and he provided explicit documentation beyond the grand generalities of the German philosopher. Prior to the so-called cell theory, Brown seems to have taken the universality of cells for granted. He was the first to assert that the nucleus is an invariable component of plant cells, and he coined the name "nucleus." He observed protoplasmic streaming. He was the substantive pioneer of paleontology having studied pollen diversity in several plant families. This list could continue.

Otherwise, Brown served the Linnaean Society as "Clerk, Librarian and Housekeeper" for some sixteen years, and he was its president four years. He was librarian and curator for Banks' unsurpassed collections of books and scientific materials. Banks willed him, in addition to a modest life income, the house and its possessions, stipulating that these then pass on his (Brown's) death to the British Museum. Brown did not wait for his own demise; rather he moved the material to the museum, there establishing a Botanical Department in the museum of which he was the keeper. That department persists to this day.

In a summation of Brown's work, Asa Gray said, "Brown delighted to rise from a special case to a high and wide generalization. He had unequaled skill in finding decisive instances . . . cautious to excess, he never suggested a theory until he had weighed all available objections to it . . . perhaps no naturalist made so few statements that had to be recalled."

Above all, perhaps, Brown was successful beyond most

botanists in achieving security that allowed quiet and tranquility in which to work, read, and think. He made the most of it. He was a true scholar of the best kind, one who well deserves Humboldt's honorary title, *Botanicorum facile princeps.*

SELECTED BIBLIOGRAPHY

Morton, A.G. 1981. R. Brown. In: History of Botanical Science, pp. 373-376. Academic Press, London.
Ramsbottom, J. 1931. Robert Brown, Botanicorum facile princeps. Proc. Linnaean Soc. London 144:17-37.
Stearn, W.T. 1970. Robert Brown. In: Dictionary of Scientific Biography. Vol. 2, pp. 516-522.

Friedrick Traugott Pursch (Frederick Pursh) (1774-1820)

Pursh, as he is known in publication, was the author of one of the early ambitious floras of North America. Of 1814, it came before **Nuttall's** *Genera* (1818) but was preceded by Michaux (1803). Thus it was a pioneer reference in its time. As such, it is yet consulted for nomenclatural reasons, and Pursh names are abundant in present-day manuals.

Pursh was born in Saxony, Germany. His formal education was minimal. His botanical education seemingly was acquired by submergence in the literature. After some preliminary horticultural experience, he obtained employment at the Royal Botanic Garden in Dresden, probably as an assistant in the garden. Also he inherited a manuscript on Dresden plants that he saw through publication. Perhaps this whetted his appetite for plants and places. He wanted to see plants other than those of Dresden. He came to the yet nascent United States in 1799, and most of the rest of his life was spent moving from one association, patron, or employer to another. He was possibly married at some time (Ewan 1981, p. 620); if so, the nature of his life suggests that it was ephemeral. What is known of him as a person seems to be that he had tartaresque features and was correspondingly rough-hewn in behavior, albeit also quick, bright, and enthusiastic. It is alleged that he was also at least a sub-alcoholic.

After initial brief residence in Maryland, Pursh was employed (about 1802-1803) by William Hamilton who had a woodland of 300 acres near Philadelphia. There he had the opportunity to meet other botanist-gardeners. Among these was Dr. B.S. Barton, who had the Lewis and Clark botanical collections and ambitious plans for a flora of North America. He took Pursh over and assigned him to collect in the eastern part of the country, which he did in 1806 and 1807. It seems that Barton intended to author a flora with Pursh's assistance. But Barton never got around to putting his proposed plans into action, and Pursh moved to New York where David Hosack, then sponsoring the Elgin Botanic Gardens, also had a spacy vision of a flora. But the financial support Hosack was expecting did not materialize.

Let me here note that I do not clearly understand the relationship of Barton, M'Mahon, and Hamilton to the Lewis and Clark material discussed at length in Ewan's (1952) paper. However, Ewan (1981, p. 608) explicitly states that Barton received the Lewis and Clark collections in 1805. (Perhaps some or most of the names of individuals mentioned in the previous paragraph may be unknown to you. But it is better than calling them X, Y and Z.)

In disgust, Pursh went to England. But he had not given up. For he took with him his specimens (or duplicates), and some Lewis and Clark material surreptitiously "scissored off" as well as his drawings and notes. And now his luck changed. He had the support of A.B. Lambert, a well-to-do English amateur naturalist and collector. Arrangements were made so that Pursh in addition to that in hand, had access to the gatherings of about forty European collectors. He wrote the *Flora Americae Septentrionalis,* and it was published in 1814.

The flora was the high-water mark; most of the rest of Pursh's life was downhill. He prepared catalogues of botanical gardens. He turned down an offer to return to the United States at Yale. He was invited to participate in an exploring expedition in Canada but that fell through after the murder of its proposed leader. He went to Montreal and tried to start on a flora of Canada. Specimens and notes were destroyed in a fire. He was without means and dependent on friends. He died. He is remembered for his flora and in the genus *Purshia* (Rosaceae).

SELECTED BIBLIOGRAPHY

Ewan, J. 1952. Frederick Pursh, 1774-1820, and his Botanical Associates. Proc. Amer. Philosophical Soc. 96(5):599-628. The basic Pursh reference—a meticulous but labyrinthine essay; with citations.
Ewan J. 1981. Frederick Pursh. In: Dictionary of Scientific Biography. Vol. 11, pp. 217-219.
This brief account is more convenient for quick reference than Dr. Ewan's research paper cited above; with citations.
Stafleu, F.A. and R.C. Cowan. 1983. Pursh, Frederick. Taxonomic Literature. Vol. 4, pp. 446-449, with citations.

René Joachin-Henri Dutrochet
(1776–1847)

Dutrochet, a physiologist, was born of a wealthy, landholding French family. Although they lost their real estate during the revolution, Dutrochet's life suggests that other fiscal resources were saved.

Briefly in the armed services, Dutrochet obtained an M.D. in 1806 and served as a medical officer in Spain for about four years. Then in 1810, he decided to throw this over and become an isolationist and research amateur, which he evidently did for the rest of his life. The fact that he had both a country home and a winter Paris residence suggests that he had francs with which to pay his bills. Apparently he never married, but lived with his mother.

Dutrochet's undertakings initially included phonetics and animal development and physiology as well as plant physiology, but some of the former were of uneven quality. His achievements in plant physiology, however, easily place him in the botanical Hall of Fame. Perhaps his major accomplishment was recognition and description of osmosis, which name he provided. Biological membranes, he said, may restrict the movement of substances in solution. But water can pass through such membranes, moving from a dilute solution of the solute to a more concentrated solution. In so doing, water develops a force resulting in pressure. This he demonstrated by devising an osmometer to measure osmotic pressure. Dutrochet found that the pressure related directly to the concentration of dissolved solute. And he cried, "Mon dieu," here is the mechanism that explains the uptake of water by roots, its subsequent movement in the plant, the maintenance of turgor, etc. Other workers rapidly picked up where he stopped.

Possibly as a spin-off from the studies of osmosis, Dutrochet investigated the mechanisms responsible for physical movement in plants (such as the so-called sensitive plant, *Mimosa*) and similarly, certain excitability responses in animals.

Dutrochet followed his contemporary, the German **Moldenhawer,** in an assertion of generalized cell theory. Treating tissues with nitric acid, he used maceration techniques and affirmed that each cell has its independent wall. He said that an organism grows because the cells both enlarge and increase in number. Some say that he was essentially the father of cell theory rather than the subsequent **Schleiden**-Swann combination. Detractors, on the other hand, note that he never clearly identified the nucleus and believe that some of his observations (as indicated by illustrations) were but optical figments resulting from deficient microscopy. Certainly the road to the cell was a long and rocky one.

Dutrochet poked into other concerns. For example, he announced that only cells with chlorophyll conduct photosynthesis. "Chlorophyll is, therefore, the essential catalyst . . . it is the stomata that make respiration in plants possible . . . it is equivalent to breathing in animals . . . mushrooms are the fruiting bodies of an underground plant body," i.e., the mycelium.

The above hither-and-yon listing of some of Dutrochet's activities suggests a bright child with new toys, feverish concentration followed by dropping one in favor of the next and the next, wherever fancy randomly struck. But no, there was, seemingly, a common thread in most of this.

Dutrochet was an avowed mechanist in a time when vitalism was the prevailing scientific and philosophical view. Vitalism held that life differs from nonlife in possessing an *élan vital* (perhaps a divine force) that sets it apart from the laws and forces that govern nonlife. And that there is no bridge between the two. Consequently, the Swedish chemist, Berzelius, coined the words organic and inorganic as contrasting terms for compounds derived from life and nonlife. Organic could be converted to inorganic but the reverse was impossible except for the agency of life.

The bridging of the unbridgeable came but stepwise. The single most significant step was the unseating of Berzelius. The chemist Wöhler (1828) heated ammonium cyanate and obtained a compound that, on testing, proved to be urea, one of the untouchable organics. Other chemists were soon falling over one another synthesizing simple organic compounds, independent of the agency of life. Perhaps the vitalism thesis was then weakened, but its primary tenets survived. Is the movement of water and mineral substances in plants dependent on an unassailable vital force? Dutrochet's studies of osmosis were directed towards providing a mechanistic explanation. So were his studies of spontaneous movement in plants. His view was that life (animals and plants) obeyed the same chemical and physical laws as those of the inanimate world. He said, "I hope the time will come when the occult and the mystic causes by which phenomena are explained will be placed by an account of the physical laws to which they are due."

And now those of us who live and work in the sheltered confines of academia know that this has come to pass. We study form and function at the atomic level. Interpretations derive from the energetics dicta of biophysics and physical chemistry. Vitalism, we say, is of the past, not worth recalling unless one has an inordinate interest in muckraking the ignorance of our ancestors.

But outside of our membrane-protected laboratories, has Dutrochet's "I hope the time will come . . ." really come? I doubt it. For example, the sales of a 1973 book by Tompkins and Bird, *The Secret Life of Plants*, it is said have far exceeded those of any botany book ever produced. *The Secret Life* relates many wonders: for example, how a *Philodendron* started a car 2½ miles away; and how the house plants of a scientist were so attuned to him that they reacted emotionally when he made love to his girlfriend 80 miles away. The book asserts that except for a few prejudiced scientists the "experts are now convinced" of the truth of the psychic life of plants.

No, this is not quite vitalism in its original form, but it is evidently the most popular botany of the day. We are sorry, Dutrochet.

SELECTED BIBLIOGRAPHY

Dutrochet, R.J.H. 1824. The structural elements of plants. In: Gabriel, M.L. and S. Fogel, eds. 1955. Great Experiments in Biology, pp. 6-9. Prentice Hall, Englewood Cliffs, N.J.
Kruta, V. 1971. R.J.H. Dutrochet. In: Dictionary of Scientific Biography. Vol. 4, pp. 263-265, with citations.
Morton, A.G. 1981. Dutrochet. In: History of Botanical Science, pp. 390-392, 410. Cambridge University Press, Cambridge.

Amos Eaton (1776–1842)

Eaton, whose father was a farmer, was born in New York. He graduated from Williams College (Massachusetts) in 1797. He then taught in country school. He passed bar examinations and subsequently made a living as a lawyer and surveyor. He wrote a book called *Art Without Science.* His legal career abruptly terminated in 1810 when he was convicted of alleged forgery, subsequently spending five years in jail and protesting his innocence to the end of his life. While incarcerated, he interested the son of the warden in botany. That adolescent was **John Torrey,** who is one of the subjects of this book.

Released from durance vile at age of forty-one, Eaton studied botany at Yale for a year and published a botanical dictionary. He became interested in science education. He also lectured anywhere and everywhere on almost anything. Apparently he was endowed with both imagination and a voluble ability to express it—"a great communicator."

According to a countdown by Merrill (1930), Eaton was married four times between 1799 and 1827 with a next generation payoff of ten children.

Subsequent to several teaching appointments and work in stratigraphic geology, Eaton became senior professor at the Rensselaer Institute in New York, which he had been instrumental in founding. He remained there the rest of his life.

Eaton's major written contribution to botany was his *Manual of Botany for the Northern States* that, between 1817 and 1840, went through eight editions. He was the first for the so-called **Gray's** *Manual* area inasmuch as his text was about thirty years prior to the first edition of Alfonso Wood's *Classbook* (1845) and the first edition of Gray's *Manual* (1848). He was the last in giving up the Linnaean classification to which he seemed fanatically devoted. On seeing Torrey's publication of Lindley's *Natural System* his reaction was apoplectic, "Since Dr. Faustus first exhibited his printed bibles in 1463, no book has probably excited such consternation and dismay as Dr. Torrey's edition of Lindley." But others did not agree with Eaton, and the natural system was in to stay.

The Linnaean system or no, Eaton's influence was not limited to his own book. For Mrs. Lincoln (Almira Hart Lincoln Phelps), his pupil and protégé, whose *Familiar Lectures in Botany* for ladies' seminaries is said, in multiple printings, to have gone through 275,000 copies.

Otherwise Eaton continued as a volatile exponent of science and education as he saw it. In a certain way he was a predecessor of the land-grant philosophy of education not just for the learned vocations, but for the people, i.e., in his words, "application of science to common purposes of life."

And Eaton's bequest to botany and science did not terminate with his death. Three of his children became science teachers. And his grandson, Daniel Cady Eaton, a descendant from second wife, Sally Cady, was a well-known paleobotanist at Yale during the latter part of the nineteenth century.

SELECTED BIBLIOGRAPHY

Ewan, J. 1969. Eaton. A Short History of Botany in the United States, pp. 39, 41. Hafner Publishing, N.Y.

Merrill, G.P. 1930. A. Eaton. In: Dictionary of American Biography. Vol. 5, pp. 605-606, with citations.

Rezneck, S. 1971. A. Eaton. In: Dictionary of Scientific Biography. Vol. 4, pp. 273-275, with citations.

Augustin Pyramus de Candolle
(1778–1841)

Most U.S. folks know that D.C. means the District of Columbia, Washington, D.C., wherein resides the benevolent management of our country. But those botanists who look at the names of plants know the abbreviation D.C. means DeCandolle, as we frequently write the name. Possibly **Linnaeus** and Candolle named more species of plants that are yet recognized by science than any other botanist.

Candolle was Swiss. While he followed the conventional course of taking a degree in medicine, his evident passion was for natural history, any and all kinds. Friendships with several botanists set his course of life. He occupied the chair of botany at Montpellier (1808-1816) and then that of natural history at Geneva where he stayed the rest of his life.

I haven't seen anything about Candolle's life as it relates to the doorsteps of his home. One supposes that he mostly just went in and out, the fate (or inclination) of most major botanists. Exception—he had a son whom I will soon mention.

Candolle's interests and work were catholic. His biggest thing was the *Prodromus Systematis Naturalis Regni Vegetabilis,* a description of the whole plant world to the species level. Unlike earlier magisterial undertakings, this included not only classification but what we would today call ecology, phytogeography, and evolution.

Candolle was the last to attempt this "everything" stunt single-handed. He failed, his death terminating the effort well before completion. The plant world had become much, much too great for the life span of any one individual. But no, he did not fail; the multivolume job was finished by his son **Alphonse de Candolle,** both as author and editor.

Only once since Candolle has a single multivolume, multiauthored effort to cover the whole plant kingdom to the species level again been attempted. That is the *Das Pflanzenreich* produced by the Engler's Berlin botanical factory, latter nineteenth, early twentieth century. Though eventually reaching to some ten feet of bookshelf, it ended far from complete.

The *Prodromus* remains a significant reference. Candolle's herbarium, containing innumerable new species, is sufficiently important in the nomenclatural sense that it has now been reproduced in microfiche and is available in the more important herbaria of the world, including that at Iowa State University. Otherwise, the *Prodromus'* role in plant classification has not been that of improving a natural classification of families and higher taxa. There were no major advances over **de Jussieu.** What Candolle did contribute was the feat of establishing infrafamilial classification in the several large families, which he studied in detail, for example, the composites, mints, and legumes.

But that is not all of A.P. de Candolle, indeed there is so much more that the best we can do here is just dull listing. Candolle's text, *Cours de botanique* (1827), followed his rewriting of **Lamarck's** *Flore Française* (1805, 1815) and his own *Théorie Élémentaire de la botanique* (1813). All of these were republished several times. He also wrote massive monographs on diverse plant families ranging from legumes to cacti. For example, his *Mémoires sur la Famille Legumineuse* (1825-1827) is really the beginning of phylogenetic interpretation of that family—all subsequently produced trace back to Candolle.

The eclectic Candolle dealt with mushrooms, the nutrition of lichens, agronomy, horticulture, pharmacology, medical botany, plant geography, the history of science, and political economy. He founded the Museum of Natural History at Geneva, renovated the botanical garden, and established the Conservatoire Botanique, which presently includes probably the world's largest herbarium (approximately 5 million specimens). He was concerned in the activities of several art societies and with philanthropic endeavors. Politics also hovered on the periphery.

Candolle's name remains with us in the epithets of several hundred species and at the generic level *Candollea* of several authors, the one surviving being *Candollea* Mirbel, a fern group and the genus *Candolleodendron* Cowan, a legume. A family named for him, the Candolleaceae of the Campanulales is now submerged under the conserved Stylidiaceae. Candolle bequeathed us his son, Alphonse de Candolle, who also enters these pages. The periodical *Candollea* is eponymic for the both of them and his grandson, Casmir. But somehow this seems like small beans. Switzerland should at least have put them on postage stamps as Sweden did for Linnaeus. But no, it was not initially possible. For it was the younger Candolle, straying far from botany, who introduced the idea of postage stamps to Switzerland.

SELECTED BIBLIOGRAPHY

Candolle, A.P. de. 1825-1827. Mémoirs sur la Famille Legumineuse, Paris. 525 pp. Facsimile reprinting, Cramer, 1967.
Candolle, A. de. 1862. Mémoires et Souvenirs de A.P. de Candolle. Geneva. Not seen.
Morton, A.G. 1981. A.P. de Candolle. In: History of Botanical Science, pp. 371-373. Academic Press, London.
Pilet, P.E. 1971. A.P. de Candolle. In: Dictionary Scientific Biography. Vol. 3, pp. 43-45, with citations.

William Jackson Hooker (1785–1865)

There is a book called *The Hookers of Kew* (Allan 1967). The plural is correct. There were two of them, our subject, W.J. Hooker and his son, **Joseph Dalton Hooker.** Between them they turned the Royal Gardens at Kew, England, from a few-acre toy of the king and affiliates into a world-famed botanical garden and botanical research institute. In addition, both were important botanists of their time.

W.J. Hooker received "a gentlemen's education," which, in his case, seemingly meant no formal education at all, at least in botany. His degrees, an LL.D from Glascow (1820) and a D.CL. from Oxford (1845), were both honorary. But as "a poet is born, not made," so are botanists. Hooker, through a gift of personality, became acquainted with most of the important European botanists and came under the sponsorship of that one-man National Science Foundation, **Sir Joseph Banks**. He was also a friend of Dawson Turner, an algologist and banker, who owned a brewery in Halesworth. Hooker married Turner's daughter (five children) and managed the brewery for some twelve years. He was not genetically endowed for the latter responsibility and was miserable. But he managed to publish several books and a monograph about mosses during this period. He was a gifted artist and prepared the plates as well as the text.

William Jackson Hooker

In 1820, he was offered, due to Banks' vigorous support, the position of professor of botany at the University of Glasgow. On paper, he was completely unfitted for the position; he had never given a lecture in his life, had no degrees, and lacked any conventional botany training. But he accepted, and unencumbered by experience, proved one of the most popular botany lecturers ever, establishing a new era at Glasgow both in teaching and research. And he rebuilt and expanded the Botanic Garden.

This led to his second professional position, director of the Kew Gardens (1841). Though he now was fifty-five years young, he jumped in with the verve of a turned-on adolescent. What he did is epitomized by an increase in the garden's holdings from about 15 acres to more than 300. And he continued as a botanical activist. Simultaneously, he maintained an avalanche of publications as a one-man book-of-the-month club in descriptive taxonomic botany.

Hooker evidently possessed an uncommon ability to project himself. Writers speak of his energy, enthusiasm, eloquence, skill in diplomacy, "commanding presence," and "charming personality." All of this, botanical and administrative, reportedly kept him busy every day from 8 A.M. until midnight. Since he had several children as well as a wife, this is possibly an exaggeration. Also Stearn's statement (1965, p. 294) that Hooker was able "to walk 60 miles a day" is certainly hyperbole or subject to a typographic error. Somewhere along here he became Sir William Hooker.

Hooker's herbarium, perhaps the largest private one then in existence, plus the Bentham herbarium became the original Kew Herbarium. Hooker contributed about 1,000 volumes for the library plus some 27,000 letters from correspondents.

The innumerable publications that he produced may be roughly classified as initially about mosses, then articles concerning anything botanical, usually with descriptions of new species (published in journals that he edited and/or established), floristic books, and revisions among the ferns. I give examples of only the latter two. Floras included the *Niger Flora* and *Flora Boreali Americana*—the latter with about 2,500 species. His fern publications, probably numbering in the hundreds, were summarized in the *Synopsis Filicum*— 75 genera, 2,252 species—prepared at the end of his life.

An observation about Hooker's monographic taxonomy is

necessary. He was the last major pre-Darwinian taxonomist. He believed in the fixity of species. His classifications tend to be typological, directed perhaps towards convenience in grouping and identification. There was evidently no thought of phylogeny as we variously interpret it today. The consequence is that the structure of many of his fern genera has been considerably modified by subsequent work. But as **F.O. Bower** has remarked, "It is better to carry out sound work on species without theorizing on their phyletic relations, than to promulgate phyletic theories without sufficient specific knowledge."

Hooker was still working on ferns a few days before his death at the age of eighty-one. He was succeeded as director of Kew by his son, J.D. Hooker, who during the latter years of his father's life had been assistant director.

SELECTED BIBLIOGRAPHY

Allan, M. 1967. The Hookers of Kew, 1785-1911. London. 273 pp.

Allan, M. 1972. W. J. Hooker. In: Dictionary of Scientific Biography. Vol. 6, pp. 492-495, with citations.

Bower, F.O. 1913. Sir William Hooker. In: Oliver, F.W., ed. Makers of British Botany, pp. 125-150. Cambridge University Press, London.

Stearn, W.T. 1965. The self-taught botanists who saved the Kew Botanic Garden. Taxon 14:293-298, with citations.

Thomas Nuttall (1786–1859)

Nuttall was born in England and died in England, but his soul was on the North American continent.

Nuttall came of a less than prosperous family. His formal schooling ended when he was fourteen. He was then apprenticed to learn the printer's trade. This he did, but simultaneously he somehow became enamored of natural science and the yet little-explored North American continent.

A self-taught naturalist and a bachelor, Nuttall departed for the United States in 1808, arriving in Philadelphia. There he became acquainted with B.S. Barton, who, although a professor of a bit of everything at the University of Pennsylvania, was really a botanist at heart. Barton helped familiarize the eager young man with the lore and literature of American botany and engaged him as a plant collector.

Arbitrarily, Nuttall's life therefrom can be divided into the following five periods: (1) four exploration and collecting expeditions (1809-1821), two in the eastern United States, one to the "Indian Country" (Arkansas Territory), and one extending to the upper reaches of the Missouri River. Some four years were wasted because he had to go to England in 1811 and the British-American War of 1812 held up his return to the United States until 1814; (2) an intermediary period (1822-1833) when Nuttall was in charge of the botanical garden at Harvard University, this vegetating, as he felt it to be, being interrupted by several leaves for collecting; (3) another trip (1834-1836), this to the Pacific Coast and then to Hawaii (the Sandwich Islands of that time); (4) 1836-1841 mostly in Philadelphia writing up collections especially for the **Torrey** and **Gray** *Flora of North America,* and preparing three volumes for an updated version of F.A. Michaux's *North American Sylva;* and (5) return to England in 1842 to take care of holdings bequeathed to him by an uncle.

Coville (1899), in a paper on Nuttall's California travels, includes interesting observations about Nuttall as of 1836 by R.H. Dana, author of *Two Years before the Mast,* an American classic about sailing ships:

> I had left him quietly seated in the chair of Botany and Ornithology in Harvard University, and the next I saw of him was strolling about San Diego beach in the sailor's pea jacket, with a wide straw hat, and barefooted, with his trousers rolled up to his knees, picking up stones and shells . . . sort of an oldish man with white hair . . . who spent all of his time in the bush and along the beach picking up flowers and shells and such truck. . . . The Pilgrim's crew christened him "Old Curious." Some of them thought he was crazy. An "old salt" explained. "This old chap knows what he's about. He ain't the child you take him for. He'll carry all these things to the college . . . and if they are better than any they have before, he'll be head of the college."

Nuttall was foremost among early American botanical explorers. But he was considerably more. In an era when recounting of the living world was primarily divided between the collectors and the closet botanists who did the writing (like Asa Gray), Nuttall was both. His *Genera of North American Plants and a Catalogue of the Species of the Year 1817* (1818) was the first of its kind prepared by an on-the-spot American botanist. He contributed mightily to Torrey and Gray's (1838-1840) *Flora of North America,* of which he perhaps should have been a junior author.

Nuttall was more than a taxonomic botanist. His *Introduction to Systematic and Physiological Botany,* prepared while at Harvard, has been said to have "partially anticipated **Schleiden's** cell theory."

Nuttall and Audubon were the first major North American ornithologists. Audubon illustrated birds and Nuttall wrote about them. His *Manual of the Ornithology of the United States and Canada* was the standard for many years.

Money, pomp, and circumstance came not to Nuttall. But is that not true for most botanists?

SELECTED BIBLIOGRAPHY

Beidleman, R.G. 1960. Some biographical sidelights on Thomas Nuttall, 1786-1859. Proc. Amer. Philosophical Soc. 104:86-100.
Coville, F.V. 1899. The botanical explorations of Thomas Nuttall in California. Proc. Biol. Soc. Washington. 13:109-121.
Graunstein, J.E. 1967. Thomas Nuttall: Naturalist. Harvard University Press, Cambridge. 481 pp., with extensive notes and citations.
Stafleu, F.A. and R.S. Cowan. 1981. T. Nuttall: Taxonomic Literature. Vol. 3, pp. 781-787, with citations.
Thomas, P.D. 1974. T. Nuttall: In Dictionary of Scientific Biography. Vol. 10, pp. 163-164, With citations.

Elias Magnus Fries (1794–1878)

Persoon was earlier, but Fries was probably the most important parent of mycological classification. He also was a pioneer lichenologist and a popularizer of science.

Fries was Swedish, a child prodigy who yet in his teens probably knew more about the fleshy fungi than anyone in the world except Persoon. His father had a doctorate in the humanities and much of the home conversation was in Latin. Young Fries studied botany and philosophy at Lund and obtained his doctorate in 1814. Immediately thereafter he became a docent, which gave him the privilege of teaching but not of salary. For several years, while writing books on lichens and fungi, he was partly supported by his indulgent father. After a series of ascending appointments at Lund, he succeeded to the professorship of botany at Upsala (1835), which post **Linnaeus** had held in the previous century. Already he was the ranking mycologist of his time.

More theoretically inclined than Persoon, Fries initially sought the divine touchstone that could be the basis of a natural classification. He was enamored by the German *Naturphilosophie* or romanticism as espoused by Okon, Nees von Esenbeck, and Goethe, which I have not the space (or perhaps satisfactory understanding) to explain. In later years though becoming more pragmatic, Fries maintained that there must be some definitive and identifiable

Elias Fries

basis for differentiating between affinity (homology) and analogy. (Such a search yet continues, the most recent romanticism being that of the cladists.) Although Fries was an evolutionist of a kind, he remained a vitalist to the extent that he found **Darwin** spiritually repugnant.

Fries much outdid Persoon in the classification and description of fungi. His early *Systema Mycologicum* (1821–1832) presented a new systematic arrangement. True, the groupings above the generic level are now primarily of historical interest. But the delineation of genera and species, deriving from his unparalleled perception of these organisms, provided the substantive background for the future, both in taxonomy and nomenclature.

Microscopes became much improved during Fries' life, but it is alleged he did not take advantage of that. Therefore he missed the spore difference between the Ascomycetes and Basidiomycetes picked up in the 1830s by **Miles Berkeley.** This may be true, but how in the world did he then study the Myxomycetes? I am not the only one to wonder about this. One writer says that "it was remarkable if not incomprehensible" how Fries' observations could have been made without a microscope.

Besides fungi and lichens, Fries also worked on a system for all plants; he regarded the Compositae "as the highest." (See R. Fries [1952], for exposition). He also published on the flowering plant flora of Sweden.

And another also. Fries wrote several semipopular books, for example, *Botanical Excursions,* three volumes. He is said to have had a "brilliant literary style, often with poetic charm." He is described (in later life) as having a tall, somewhat bent figure, silvery-white hair, and sparkling eyes.

Valid publication for names of fungi presently dates to 1821 with Fries' *Systema Mycologicum* and *Elenchus Fungorum* except for the Myxomycetes and for those groups referred to the earlier Persoon (the Uredinales, Gasteromycetes, and Ustilaginales). By providing a descriptive classification structure, Fries not only opened the doors for nomenclature but for successors such as **de Bary,** who now turned to life cycles, morphology, and physiology.

R. Fries (1952) says of his grandfather that "like all real geniuses, he was primarily everything good and benevolent." (As blessings be, that may have been true of Fries; but as a generalization it is sadly lacking.) In his time, Fries achieved the status of "Grand Old Man" of Swedish natural science. And I am pleased that he also had "poetic charm."

SELECTED BIBLIOGRAPHY

Eriksson, G. 1972. Elias Fries. In: Dictionary of Scientific Biography. Vol. 5, pp. 190–192, with citations.
Fries, R.E. 1952. Elias Fries. Lindroth, S., ed. Swedish Men of Science, 1650–1950. Swedish Institute.
English translation by Burnett Anderson, pp. 178–185.
Stafleu, F.A. and R.S. Cowan. 1976. Taxonomic Literature. Vol. 1, pp. 878–888. Extensive bibliography and listing of publications.

John Stevens Henslow (1796–1861)

The fact that Henslow, a botanist, was the maker of **Darwin** is sufficient to include him in this series. But beyond that he was "a man of rare character and singularly extensive acquirement in all branches of natural history."

Henslow, born in Kent, England, was one of eleven children. He went to Cambridge. An enthusiastic naturalist, his expertise and knowledge was so quickly evident that he was appointed professor of mineralogy at Cambridge in 1826 and of botany the next year, the latter continuing to the end of his days. But he had two professional lives. Trained also in theology, he had appointments successively as a curate and vicar, finally coming to the rectorship in Hitcham, Suffolk, in 1839, which he maintained for life. Evidently he resided in Hitcham, only going to Cambridge for relatively short periods for his lectures.

Henslow married in 1833; there were five children including a daughter who married the famed botanist, **Joseph Hooker.**

Although little is said about Henslow's personality, one receives the impression that he was a much-loved man. He even looks like a pleasant fellow in the only picture I have seen, rather than the austere and forbidding aspect presented by most nineteenth-century botanical portraits.

As a friend of plants, Henslow, beyond the name of each, wanted to know its personality, how it grew, why was it there, what habitats did it favor, and how was it distributed. Humboldt, the famous plant explorer, was his god. One author says Henslow was a "genuine ecologist without knowing it." Another says he was the forerunner of **Schimper.** His teaching was a mixture of lecturing and demonstrations, the latter presented to the students as problems so that they would learn by thinking, explaining, and doing. The teaching of botany flourished at Cambridge as never before. Henslow's publishing included, among perhaps fifty titles, a *Descriptive and Physiological Botany* (1836) and a *Catalogue of British Plants* (1829 and 1835). A *Flora of Suffolk* was incorrectly attributed to him.

In his other life in Hitcham, I suppose he had some churchly duties, but no one says anything about them. It is obvious, however, that he kept busy. He was a one-man agricultural extension service. He wrote a sequence of newsletters aimed to help farmers mix their art with science, the better to make a living. He helped Lawes of the Rothamsted Experiment Station locate a source for phosphate fertilizers. He held classes for children. He gave them a checklist of the names of flowers and the time of blooming; they were to bring samples of these in succession and put them in water containers in the laboratory. He took them on field trips. During the time of the potato famines, 1847-49, he devised a way of recovering the starch from rotten potatoes. But he struck out one time. The farmers thought their wheat rust came from the barberry, and Henslow could not convince them that two kinds of fungi were concerned. The farmers were more nearly right than the professor.

We must switch back to the Cambridge career for the principal story. The young Darwin, at the wish of his family, had tried both the medical profession and theology. He liked neither. A depressed young man, he had the extraordinary good fortune to come under the wing of Henslow, who introduced him both to botany and geology. He brought Darwin and Sedgewick, a geologist, together, and Sedgewick took Darwin on rock exploring expeditions in their part of England. Henslow's master stroke was his recommendation that Darwin be the naturalist on the H.M.S. Beagle. "Where the Edinburgh medicos, and the Cambridge theologians had signally failed, Henslow and Nature showed him the way." And this was, as

Darwin subsequently said, the starting point of his life.
Is a summary eulogy of Henslow necessary? I think not.

SELECTED BIBLIOGRAPHY

Boulger, G.S. 1891. J. S. Henslow. In: Dictionary of National Biography (Great Britian). Vol. 26, pp. 135-136.
Henslow, G. 1913. J. S. Henslow. In: Oliver, F.W. ed., Makers of British Botany, pp. 150-163. Cambridge University Press, London.
Mathews, M.V. 1972. J. S. Henslow. In: Dictionary of Scientific Biography. Vol. 6, pp. 228-229.

John Torrey (1796–1873)

Torrey, never a professional botanist, was for some fifty years the most important amateur seer of mid-nineteenth century American descriptive and floristic botany. He was second only to his protégé, the overpowering **Asa Gray,** whom Torrey first discovered and then firmly established in the initial role of a Torrey collaborator.

Torrey was born in New York City, the second of ten children. His father for some period was a warden or fiscal agent for the Newgate state prison, which then included among those incarcerated botanist **Amos Eaton,** put away for some alleged irregular dealing in real estate. (Eaton persistently insisted his innocence.) Young Torrey became acquainted with Eaton who converted him to botany, though not in the sense of making a living therefrom.

So Torrey was trained as an M.D. He practiced briefly. Then he took up teaching, primarily chemistry, natural history, and geology. For some years he held concurrent positions at the College of New Jersey (subsequently Princeton) and the New York College of Physicians and Surgeons. He ended his career at Columbia where he was given a house in return for his herbarium and personal library (Dupree 1976).

Torrey married in 1824 and had three daughters and a son. Gray, who lived in New York prior to moving to Harvard, was a fre-

John Torrey

quent visitor in the congenial Torrey home, and it was assumed among friends that he would marry a Torrey girl—which one was the only question. However Gray dallied until age thirty-eight when he eventually chose (or was chosen?) elsewhere.

Torrey's accomplishments as a botanist were presumably in his spare time. He and Gray (the T. & G. of numerous author citations for new species) initiated a *Flora of North America* (1838-1843). It was never completed because both writers became inundated with material from diverse collectors and exploring expeditions, and these had to be taken care of before the *Flora* could be picked up again.

Perhaps Torrey's most satisfying work was his *Flora of New York*, two volumes, 1843, a model for its time. But afterwards, in fact for most of the ensuing years of his life, he and Gray, in part or entirely, were writing up the botany for the several government exploring expeditions of the western part of a pioneer country trying to find out what it had and where its boundaries were. There were for example, the Fremont expeditions (1842) and 1843-1844, Emory's of 1846-1847, and the Mexican Boundary Survey, which had a sequence of leaders and botanists. And beyond the continental limits of the United States came the South Pacific Wilkes Exploring Expedition, perhaps an attempt to compete with **Darwin's** Beagle foray. These and others are detailed by Rodgers (1942). The route to publication of some of these are horror stories of bureaucratic dalliance. Torrey joined none of these expeditions in the field. His role was that of "writing up" the botany collections and in some instances pulling strings concerning the participating field botanists. Only in later life did he hold several governmental appointments that enabled him to travel the West and for the first time see many plants formerly known to him only from herbarium sheets.

With one exception, Torrey, unlike Gray, dealt with none of the burning issues of biology like the Darwinian proposition in which the latter was so active. The exception was the editing and publication of an American edition of Lindley's *Natural System of Botany* (1830), which brought with it the issue of the Linnaean vs. the natural classification. At this time, the Linnaean classification still held sway with many American botanists, and the Lindley-

Torrey production was greeted with mixed response. Amos Eaton, an author of botany manuals and protagonist of **Linnaeus,** damned the book with some violence. (See essay about Eaton.) But the natural system rapidly took over and the Linnaean sexual arrangement was retired to history.

Torrey may not have dealt with the "burning issues" like Gray, but he was involved in others. For example, he was enzymatic in establishment of the U.S. National Herbarium. His own herbarium became the Columbia University Herbarium, which in turn was the nucleus for that of New York Botanical Garden at the end of the century.

John Torrey was a gifted, pragmatic systematist and a superb human being. All of several references seen are unanimous about his kindly personality, geniality, and helpfulness to others. In addition to conventional honors, the genus *Torreya* (a rare Florida Gymnosperm) is named for him as are species in diverse genera. And the *Bulletin of the Torrey Botanical Club,* now more than 100 young, continues as our most venerable journal.

SELECTED BIBLIOGRAPHY

Barnhart, J.H. 1936. Torrey. In: Dictionary of American Biography. Vol. 18, pp. 596–598, with citations.
Dupree, A.H. 1976. Torrey. In: Dictionary of Scientific Biography. Vol. 13, pp. 432–433, with citations.
Robbins, C.C. 1968. John Torrey: his life and times. Bull. Torr. Bot. Club. 95:519–645, with citations.
Rodgers, A.D. III. 1942. John Torrey. A Story of North American Botany. Princeton University Press. 352 pp.

George Bentham (1800–1884)

Bentham, premier systematic botanist of the nineteenth century, came of a wealthy and aristocratic English family. His father, Samuel Bentham, was inspector general of the navy and his uncle, Jeremy Bentham, was a noted moral-ethical philosopher and economist.

George Bentham's education, mostly by private tutors, was continuously interrupted by movements of the family from country to country where his father's duties called him. Owing both to necessity and natural gifts, Bentham early acquired reasonable proficiency in some fourteen languages. His subsequent formal training was in law. For several years he managed the family estates and served as Man Friday for Uncle Jeremy. During this period, along with music (he was said to be an excellent pianist), philosophy, and mathematics, botany was a hobby. His first botanical work was a catalogue of the plants of the Pyrenees. He also wrote a book on logic. He started a diary that continued until shortly before his death, nearly half a century later. He married in 1833—there were no children.

Bentham, the prince of systematic botanists, was a research loner whose only joint publication was his terminal work, the Bentham and **Hooker** *Genera Plantarum.* His primary participation in human affairs was that of the role of president of the Linnaean

Society for thirteen years beginning in 1861 (his diary reveals that this to him was an exasperating experience). He stayed away from the noise of conflict; for example, he took no part in the Darwinian imbroglio that involved many of his friends. On the other hand he gave much individual help to botanists who sought his counsel; for example, he spent several months working with **Asa Gray** who was faced with the identification of plants of the U.S. trans-Pacific exploring (Wilkes) expedition. He is pictured as a tall, austere man, clean shaven, most of his life with a shock of black hair and a "hawk-like" face.

Following the death both of his father and uncle (1831 and 1832), Bentham inherited holdings sufficient for life and retired to botany. He was immediately writing vast monographs and contributing to **de Candolle's** *Prodromus*. He assembled a private library and herbarium, the latter coming to 100,000 specimens. As a consequence of his friendship with Sir Joseph Dalton Hooker, director of the Royal Botanical Gardens at Kew, in 1854 he moved his working quarters from his house to Kew. The combination of his holdings and those of Hooker's represented the genesis of the presently unexcelled library and the now 5-million-specimen Kew Herbarium. And there Bentham remained the remainder of his life, unsalaried, of course, living in a world of plant specimens. Certain tradition has it that his coach would arrive precisely at 10:00 A.M. and leave at 4:00 P.M. so that one could set a watch by his movements. But I have also seen a note somewhere that he usually worked longer hours; who now knows.

Of Bentham's activity, continued about fifty years, I hazard the opinion that if one jointly evaluated the quality and quantity of individual production, he was the outstanding systematic botanist of the nineteenth century. (In volume, of course, he could not by himself compete with the subsequent **Engler** botanical factory in Berlin.) By quality I mean the combination of his jurisprudential judgment and intuition in considering both the "natural affinities" of the major orders of plants and in delineation of individual species, often from just a few specimens. His botany was both phylogenetic and descriptive. Subsequent botanists retreading Bentham's ground with much more material and using more sophisticated methodology commonly have only confirmed

Bentham's conclusions. His species were "good" species, or, if there were exceptions, his batting average likely exceeded that of any other botanist of the past.

Now some samples from the production line. Following an abundance of preliminary papers, Bentham published revisions of large portions of the immense families of the mints, legumes, and composites. At least for the legumes and mints, he contributed more than anyone during or prior to his time. He prepared family treatments for major multiauthored works as Candolle's *Prodromus* and Martius' *Flora Brasiliensis*. He identified plants from collections of various botanists and government exploring expeditions that sampled the extremes of the then mighty British empire. He reported the results, including the description of innumerable new species, in his *Plantae Hartwegianae* and *Botany of the Voyage of the H.M.S. Sulphur*. He wrote a *British Flora* (1858); his *Flora of Hong Kong* (1861) was dwarfed by the *Flora Australiensis,* seven volumes done in about eight years, published 1863-1878. One hundred years later it, remains a major reference. (I used it in the 1960s in determining the numerous species of Australian *Acacia* planted in California. I hasten to add that since then Australian botanists have made substantial progress in treating *Acacia* and other groups and areas in their country.)

Stafleu (1966) has provided information about the nineteenth-century Kew effort to produce or stimulate the preparation of colonial floras. And he explains a matter that has puzzled me. It is why *Flora Australiensis* lacked the participation of the eminently qualified Australian botanist, Baron von Mueller, and why, on the other hand, the work has sometimes been cited as Bentham and Mueller. In summary, it was judged that to provide an authoritative flora, it was necessary to combine the Mueller holdings in Australia with those at Kew and accessible in other European herbaria. J.D. Hooker, director of Kew, tried to get Mueller to come to England with his materials ("life is short and books are long"). But Mueller, for reasons now unknown, was unable to do so and finally acceded to loaning his collections, publications, and notes to Kew for Bentham's use—this must have been hard on Mueller. The citation of authorship "Bentham and Mueller" arises from the title page listing of Bentham, "assisted by Ferdinand Mueller." Mueller did indeed "assist" with the materials, but the authorship was exclusively Bentham. Stafleu reminds us also that this undertaking of Bentham's plus the *Genera Plantarum* was done between his seventieth and eighty-third year.

Bentham's production of *Genera Plantarum* included Joseph Hooker as junior author, his (Bentham's) only collaborative undertaking. *Genera Plantarum* constitutes a descriptive listing of *all* families and genera of Angiosperms and Gymnosperms known to the authors. These were arranged in an affinity sequence considerably modified from the earlier multiauthored Candolle *Prodromus*. Issued in several parts that were assembled into three volumes, published 1862-1883, the total treats 7,565 genera in 6,617 pages. Within each of the larger families, the genera are grouped into tribes and subordinate divisions, i.e., a conspectus summarizing the structure of the family. The generic descriptions are models of concise excellence. Distribution and critical notes concerning each generic entry were also briefly presented. Although much dated (and written in Latin), the work yet remains a major reference and is a monumental representative of the best in systematic botany of the nineteenth century.

Despite or perhaps beyond his aristrocratic aloofness, Bentham was evidently a gentle man—not always true of those of eminence. In an obituary in *Nature*, Sir Joseph Hooker said, "Of his amiable disposition and his sterling qualities of head and heart, it is impossible to speak too highly; though cold in manner and excessively shy in disposition, he was the kindest of helpmates."

SELECTED BIBLIOGRAPHY

Bentham, G. 1884. On the joint and separate work of the authors of Bentham and Hooker's Genera Plantarum. J. Linn. Soc. London 20:304-308.
    Bentham explains "why" for "the only joint work in which I have ever been engaged."
Jackson, B.D. 1906. George Bentham. J.M. Dent and Co., London. 292 pp.
Stafleu, F.A. 1966. Bentham and Hooker. Taxon 15:37-39.
    About the monumental Genera Plantarum authored by Bentham and Hooker.
Stafleu, F.A. 1967. The Flora Australiensis. Taxon 16:538-542.
Taylor, G. 1970. Bentham. In: Dictionary of Scientific Biography. Vol. 1, pp. 614-615, with citations.

Adolphe-Theodore Brongniart
(1801–1876)

Brongniart, a several-faceted French botanist, was the first important paleobotanist, or "the founder of palaeophytology" of one writer.

A.-T. Brongniart was the son of Alexander Brongniart, geologist and paleontologist. Information about his education in anything seen is nil. His only appointments seem to be those of professor of botany with the Museum d'Histoire Naturelle in Paris in 1833 and the next year with the Academy of Science.

What one does see is that by 1828 (age twenty-seven) he was apparently the authority of his time on fossil plants. First there is his *Histoire des végétaux fossiles,* largely descriptive, published in fascicles (1828-1838) but never completed. Then the *Prodromus* (1828), which is of theory and interpretation. Brongniart divided the plants of the past into four periods based on the groups sequentially dominant in the fossil record. He proposed a classification of plants (including those extinct) into some six groups, being the first to place the Gymnosperms as a class among the Phanerogams (higher plants with evident sexual reproduction), there parallel to the Monocots and Dicots of the Angiosperms.

These publications alone support the contention that, to the extent that any single person can be named as the father or founder of a discipline, Brongniart was that for paleobotany. Not only did he provide descriptions and classification, but within the constraints of his time he developed basic concepts of interpretation. Several of these were of geological nature (obviously his training and interest included geology). For example, he applied the stratigraphic principle to fossil plants, i.e., that certain kinds of plants should be characteristic of certain strata wherever these were found.

There is some mystery about the termination of his uncompleted *Histoire*. It was published (1828-1838) up to volume two and then it ends on page 72 "in the middle of a sentence" (Stafleu 1966). Something seemingly interrupted him. Perhaps the mail came one morning. He stopped to open it and saw notification that he was wandering too far from appropriate biblical tradition, with a promise of impending doom if he didn't mend his ways.

True Brongniart had stayed mostly within the approved trodden path, and indeed he was said to have been a disciple of the famous French paleontologist, Cuvier, whose speciality was the fossil record of fish. Cuvier was a strict fundamentalist who believed neither in the mutability of species (a nasty idea at this time) nor in the continuity of the life of the past. Rather, his position was that there had been several creations followed by catastrophic extinctions (equivalent to Noah's flood?), each being succeeded by another creation. Thus, as an apocalyptist, he was able to study the various kinds of fossil fish without colliding with biblical doctrine.

Keep in mind that this was before **Darwin,** and that the famous and flamboyant **Comte du Buffon** previously had to recant some heretical statements about progressive changes in species. Perhaps Brongniart was happy to ride in Cuvier's protective shadow. But did he buy Cuvier whole hog? I certainly doubt it. As Cuvier, Brongniart recognized extinction. He attributed extinction to changes in climate, which presumably implies gradualism rather than catastrophe. And of fixity of species? Morton (1981) quotes an interesting sentence from the Brongniart *Histoire* (1828), a part of which follows, "By studying ancient floras ... we may be able to give greater probability to *theorie at present considered to be simple hypothesis.*" (Italics mine.) Was this a hint about the possibility of mutability of species, i.e., evolution?

In any event, except for a final synthesis (*Végétaux fossiles,* 1848) that contained no new theoretical lights, Brongniart stopped and became a botanist of the present. He did good work as a physiologist and taxonomist, but he was not a pioneer. He was especially interested in the flora of New Caledonia from whence he described numerous genera and species. The baton of paleobotany went to the English.

The leguminous genus *Brongniartia* Kunth (1824) and the tribe Brongniartiae are named for our subject, as well as, independently, *Brongniartia* Blume (1825) of the Monimiaceae (a tropical family of the Laurals). And also there is *Adolphia* (Brongniart's first name) Meisner of the Rhamnaceae. I have worked some with the legume *Brongniartia,* which is largely a Mexican genus; the species I know are ugly desert shrubs. But such is fame.

SELECTED BIBLIOGRAPHY

Leroy, J-F. 1970. Brongniart. In: Dictionary of Scientific Biography. Vol. 2, pp. 491-493, with citations.
Morton, A.G. 1981. Brongniart. In: History of Botanical Science, p. 414. Academic Press, London.
Stafleu, F.A. 1966. Brongniart's Histoire des végétaux fossiles. Taxon 15:320-324.
    Among references here cited and others seen, this is the most critically informative.
Stafleu, F.A. and R.S. Cowan. 1976. Brongniart. In: Taxonomic Literature. Vol. 1, pp. 352-355, with citations.

Miles Joseph Berkeley (1803–1889)

Berkeley, an English clergyman, was among the most distinguished mycologists of the nineteenth century. He went to Cambridge where the botanist **Henslow** interested him in biology. Henslow, you may remember, was the one who also started **Darwin** on the way. But biology was not Berkeley's survival profession. Within a year after his graduation (1825), he was ordained in the English church, sequentially a curate, a "perpetual curate" (well, almost perpetual, thirty-five years), and finally a vicar. He married in 1830; the consequence was fifteen children. To supplement his meager church income, he (and certainly also his anonymous wife) ran a small boarding school for boys.

Available references all seemingly copy from one source about Berkeley's personal mien. With no further information, I can only repeat: "Berkeley was a man of splendid presence and great refinement . . . a tall, commanding figure, grand head of flowing white hair."

Since Berkeley's mycology was a product of his leisure time, I can only suggest that his perpetual curate duties must have been negligible or that, sequestered with his fungi, he grossly neglected them.

But onward to science! Berkeley started with marine algae. Thereafter he was mostly a mycologist with phytopathological overtones. He prepared the fungi for J.E. Smith's *English Flora* (ca. 1830), 155 fungus genera, 1,360 species, many of them new. This alone placed him as the foremost British mycologist. And he continued nearly every year (1837-1883) to publish on the fungi of his country, often jointly with his friend, C.E. Broome.

Berkeley came to the attention of **W.J. Hooker,** director of Kew. Berkeley was soon the Kew authority for the fungi of the world; all exotic collections passed through his hands. He described species from everywhere. This arrangement lasted to the end of Berkeley's life and through three directors of Kew. All in all, he is said to have written on several thousand species in some 400 papers (presumably no page charges). And there were two books, *Introduction to Cryptogamic Botany* (1857) and *Outlines of British Mycology* (1860). And according to Taylor (1970), he read proofs on the Bentham/Hooker *Genera Plantarum* (three volumes, 7,565 genera, 6,617 pages.

Berkeley was the first clearly to define the Basidomycetes on the basis of the four externally produced spores. Beyond his primary identification-description thrust, he wrote about the distribution of fungi. His two primary points of view, it has been said, were that (1) adapted fungi are more cosmopolitan than higher plants, and (2) their distribution is secondary to that of flowering higher plants, their hosts, which they necessarily follow. This simplistic categorization may not do Berkeley justice, inasmuch as it does not differentiate between saprophytes (nourishment from dead organic matter, e.g., bread mold) and parasites whose range must be limited to that of their host or hosts.

Berkeley was also a pioneer plant pathologist. I hesitate to say that "he was the originator and founder of plant pathology" as Massee (1913) suggests, but "the first noted British pathologist" seems reasonable. Berkeley's study of the potato blight is most commonly mentioned. He demonstrated that the casual agent was *Phytophthora infestans* and worked out its life history. He maintained a running series of popular articles on plant pathology in the *Gardeners Chronicle* for several years. There seems little doubt that, as no one else of his day, he was aware of the economic significance of plant pathology.

In another direction, Berkeley wandered, albeit probably

briefly, into the world of green plants, namely bryology, publishing a *Handbook of British Mosses* in 1863. Evidently it was used, for there was a second edition in 1895. Three fungus genera plus one of the bacteria have been named for Berkeley: *Berkeleya, Berkelella, Berkeleyna,* and *Berkleasium.*

At retirement, Berkeley received his first government support, a pension of 100 pounds a year. Hallelujahs to him as a gifted and dedicated amateur professional. But I still wonder how he, his family, and parishioners survived.

SELECTED BIBLIOGRAPHY

Massee, G. 1913. Miles Joseph Berkeley. In F.W. Oliver (ed.). Makers of British Botany, pp. 224-232. Cambridge University Press, London.
Stafleu, F.A. and R.S. Cowan. 1976. Rev. Miles Joseph Berkeley. In: Taxonomic Literature. Vol. 1, pp. 192-195, with citations.
Taylor, G. 1970. Miles Joseph Berkeley. In: Dictionary of Scientific Biography. Vol. 2, pp. 18-19, with citations.
Whetzel, H.H. 1918. Miles Joseph Berkeley. In: An Outline of the History of Phytopathology, pp. 55-57, with citations. W.B. Saunders Co.

Jacob Mathias Schleiden (1804–1881)

That cells are the universal building blocks of which all animals and plants are constituted was "discovered" in 1838. Schleiden, a botanist, and Schwann, a zoologist, got together, compared notes, found they had the same ideas and published, "Voilà and how, how; it is the cell!" At least that is somewhat as the Schleiden-Schwann assurance is reiterated in elementary botany and biology books to date. Nonsense. The development of rational cell theory was an incremental process. Schleiden/Swann were just a watering place along the way. True their views represent a significant step in its avowed application to plants and animals alike. But it was an adolescent thesis, nearer the beginning than the end.

But to our subject otherwise. Schleiden, German, came of a middle-class family. His father was a physician. He studied law, getting a doctorate in 1827. Then he practiced. He didn't like it. So he retreaded and obtained a Ph.D. in botany in 1839. Thereafter he lectured and wrote. He was a dazzling popularizer who drew students by the droves. He had a polemical and combative personality, which he used in insulting and ridiculing all others. Evidently tangling with his colleagues, he was constantly on the move from one institution to another. He wrote a widely used textbook *(Grundzuge der Wissenschaftlichen Botanik)* that was translated into several languages. He vigorously attacked not only the philosophers and botanists who espoused the so-called German romantic (ultimate causes) philosophy, but others who came to his attention. One of those whom Schleiden devastated responded with a gentle book, *Professor Schleiden and the Moon.*

I have seen nothing about Schleiden's nonbotanical life, if any. His personality seems sufficiently revealed by his professional attitudes.

Schleiden, yet young, a year from his degree, first comes to notice in the Schleiden/Schwann combination. But perhaps it was an 1838 coincidence rather than a real combination. Mature cell theory required assurance that cells start out as those in parenchymatous tissue with a discernible nucleus, and that new cells arise only from pre-existing ones (*omnia cellula e cellula*). And what of the role of the nucleus?

Cells had been seen in plants since or before the time (1665) of **Hooke's** *Micrographia,* but scarcely so in animals because of their functional modifications. As it is related, Schwann settled the matter for Schleiden by showing him plant-like cells with nuclei in the tails of tadpoles.

All right, but how do existing cells form new ones? Schleiden said those *de nouveau* come internally from the nucleus (cytoblast in his terminology) of pre-existing cells through a sequence described in recent literature both by Morton (1981) and Klein (1975). (But the interpretation of what Schleiden meant somewhat differs between these two authors.) How does the new cell get free from the old; how does the original cell replace its nucleus; what about cell wall formation?

It turns out that the Schleiden microscopic studies were based on free cell formation in the embryo sac of angiosperms. There cell division without wall formation can be easily seen but it is a rather special case. Authors immediately following Schleiden, especially **von Mohl** and **Nägeli,** quickly saw that cell wall formation had to be considered. The thesis was considerably dismantled.

But cell theory was just an early incident in Schleiden's career. Subsequently, it was really his textbook, *Die Pflanzen und ihr Leben,* that put him in the big time (i.e., in the micro world of botany). It was indeed the first of the great German texts. Going through several editions and translations, it changed the face of botany.

Jacob Schleiden

Microscopic examination was exalted, and systematics became the weak tail rather than the dog; inductive methodology was hallowed and philosophical theorizing rejected.

In context now of the twentieth century, one must overlook Schleiden's inappropriate attacks on earlier experimenters (including such individuals as **Stephen Hales** and **Joseph Priestley**) if we want to give him credit for the positive influence of his textbook. Beyond this, as his years went on, Schleiden continued to present his dogma on science in popular lectures. Some were assembled and printed in books. And he wandered increasingly to numerous topics beyond science, for example, concerning the troubles of the Jews in the Middle Ages. (Likely he witnessed ethnic prejudices in his time.)

To the end of his life, Schleiden was involved in harsh polemics with others. The author L. Errea (Klein 1975) aphorized his career: "As a popularizer he was a model; as a scientist an initiator." And I might again refer to *Professor Schleiden and the Moon*. Perhaps it may serve as the best identification of: the communication of various moons with sufficient persuasive vigor that they seemingly became realities. And some were.

SELECTED BIBLIOGRAPHY

Coleman, W. 1971. Schleiden and others concerning "cell theory." In: Biology in the Nineteenth Century, pp. 23-31. John Wiley, N.Y.

Klein, M. 1975. Schleiden. In: Dictionary of Scientific Biography. Vol. 12, pp. 173-176.

Morton, A.G. 1981. Schleiden. In: History of Botanical Science, pp. 377-384. Academic Press, London.

Hugo von Mohl (1805–1872)

Von Mohl was an unusual botanist in that by account he lived a "gentle and prosperous life." Coming of a solid middle-class family, he had a "happy childhood," a satisfying educational career, a "personal life without difficulties" (he never married), and was much recognized and awarded during his lifetime.

Professor of botany at Tubingen, Germany, von Mohl was a skilled microscopist. He wrote a book about his art as well as one entitled *Die Vegetablische Zelle,* a fundamental reference in its time. His work consisted mostly in accurately discerning what others had grossly misinterpreted or had not seen before. His numerous findings were mostly empirical observations with a minimum of theory. Among these, I note two of significance.

No one previously had provided a clear account of the monocot vascular bundle. Von Mohl did so, concluding that the tissues are the same as in dicots; the difference is only in that they are arranged differently. Subsequent botanists concurred and so proclaimed in botany textbooks. Because xylem and phloem arrangement has now been inertly observed by thousands of students from prepared slides, one wonders if the whole business was not a simple matter previously mangled by idiots. Granted that pioneer botanists were sometimes endowed with an overabundant imagination, one

Hugo von Mohl

thinks not. It is necessary to remember that in the early part of the nineteenth century, biologists were working before the development of adequate microtechnique procedure, i.e., microtome sections and staining, and what was seen was often a function of methodology.

Von Mohl picked up on the cell that **Schleiden** had left in disturbed condition with mysterious internal budding to produce more of itself. True, von Mohl did not get around to mitosis, which came shortly after his time. Rather, he provided the alternative to Schleiden's budding, a careful description of the formation of the new cell wall coincident to cell division, the middle lamella (that term was not used) followed by the primary and successive secondary layers on each side. He also was concerned with the cell contents. He used the word *protoplasm* for the living part of the cell as distinguished from the vacuole, cell membrane, and wall. He defined the cell in the sense of the vital protoplast unit, the wall being just the inert container of the life within.

I repeat that von Mohl was not a speculator. In all of his work (the preceeding being only examples), he stayed within the careful limits of observation and attempted no grandiose biological theories as had some of his predecessors. It goes without saying that he was not a compatriot of the German romanticism (*Naturphilosophie*) much in vogue during his time.

At Tubingen, von Mohl espoused the creation of a Faculty of Sciences, said to be the first of its kind in the Germany of then. He was instrumental also in the founding of the *Botanische Zeitung*, one of the most long-lived of the major botanical journals of all time. He died peacefully in his sleep.

SELECTED BIBLIOGRAPHY

Klein, M. 1974. H. von Mohl In: Dictionary, Scientific Biography. Vol. 9, pp. 441-442, with citations.

Morton, A.G. 1981. von Mohl. In: History of Botanical Science, pp. 386-388. Academic Press, London.

Sachs, J. 1875. von Mohl. In: Geschichte der Botanik. English translation by Garnsey, H.E.F. and I.B. Balfour, 1890, pp. 291-310. Clarendon Press, Oxford.

Alphonse de Candolle (1806–1893)

The second Candolle was the son of **A.P. de Candolle**, an earlier inhabitant of these pages. He equally deserves posterior acclaim.

Alphonse was born in Paris and grew up in Montpellier. Although his undergraduate degree was in science (1825), he followed with a Doctorate of Law in 1829. This peregrination to the legal world not withstanding, he succeeded his father as professor of botany and director of the Botanical Garden at Geneva, Switzerland, in 1835. Also he then inherited responsibility for completing the monstrous, multivolume *Prodromus Systematis Naturalis Regni Vegetabilis* initiated by the elder Candolle. Simultaneously he was active in politics and public affairs; for example, he introduced postage stamps into Switzerland. He was a vigorous participant in various learned societies from whom he reaped diverse honors. He married in 1832. There was a son, Casmir, who also became a botanist and later collaborated with his father.

Like the prior Augustus P. Candolle, Alphonse was into a bit of everything. His voluminous botanical output was approximately of four kinds (1) major books, for example, a botany text soon translated into several languages, a text on plant geography said by Pilet (1971) to be "still the key work of phytogeography" (certainly a pioneering work; scarcely at present a key work); (2) numerous plant

Alphonse de Candolle

family monographs; (3) management of completion of his father's *Prodromus*, author in part, editor, and business manager; and (4) miscellaneous papers on diverse botanical topics. He also wrote a history of science. A scan of his publications shows sociology, medicine, meteorology, politics, and problems of scientific societies.

I am familiar with two of his productions. The first is the *Lois de la nomenclature botanique* (1867), of which he was the editor and to a considerable extent the author. This was the first of our Codes of Botanical Nomenclature. Candolle was president of an International Botanical Congress in 1866 called to consider the chaos into which plant nomenclature had fallen. It is true that there was a general gentlemen's agreement about priority, i.e., that the first published name for a species or genus was that to be used. But not all authors were gentlemen, and there was no consensus concerning the numerous factors defining publication (e.g., the use of an old name in a new combination) and when priority should start—with **Linneaus** or **Theophrastus?** The congress was followed by another international meeting the next year for which Candolle by then had prepared the *Lois de la nomenclature botanique*. Subsequent editions of ICBN (International Code of Botanical Nomenclature) have much evolved in detail, but the fundamentals were in Candolle's *Lois*.

And there is the *Origine des plantes cultivée* (1882), which has seen translation into several languages. I have used the book extensively and otherwise read it. Employing the available evidence from a wide variety of sources, Candolle reasonably identified the geographic origin of most agronomic, horticultural, and fruit crop plants. It is a sane, careful treatment, and modern data substantiate most of the conclusions. Despite subsequent contributions by Vavilov, Sauer, and Merrill, it has really been replaced only by Jack Harlan's recent *Crops and Man* (1975).

Candolle noted that cultivated species of the New and the Old Worlds were apparently mutually exclusive prior to Columbus. He suggested, contrary to the then dominant diffusionism thesis, that this could well indicate that the agriculture and derivative civilizations in the Americas were entirely independent of those in the Old World, and were not established by any wandering lost tribes of Israel or whomsoever. With somewhat more data and without the

Candolle caution, E.D. Merrill vigorously espoused this thesis in 1938.

I guess Candolle, an eclectic organizer and critical analyst, really built no entirely new houses on the plant science subdivision. This is suggested by the fact that Morton (1981) barely mentions him. But he did remodeling, installed much plumbing, and otherwise updated structures standing or started.

SELECTED BIBLIOGRAPHY

Candolle, A. de. 1886, 2d ed. Origine des plantes cultivées.
    English translation and reprint, 1959, as Origin of Cultivated Plants. Hafner, N.Y. 468 pp.
Christ, H. 1893. Notice biographique sur Alphonse de Candolle. Bull. Herb. Boiss. 1:203-234.
    In French.
Morton, A.G. 1981. History of Botanical Science. Academic Press, London. 474 pp.
Pilet, P.E. 1971. Alphonse de Candolle. In: Dictionary of Scientific Biography. Vol. 3, pp. 42-43.

Alvan (Alvin) Wentworth Chapman
(1809-1899)

The first regional flora in the United States were naturally of the northeastern states, the first settled and the most developed in a yet expanding country. The productions were by professionals, in the sense at least that they were college or seminary teachers. Excluding the **Torrey** and **Gray** pioneer and incomplete *Flora of North America* (1838-1840), the first to dare beyond the northeast states boundaries was Chapman, a practicing physician who was strictly an amateur. His *Flora of the Southern United States* appeared in 1860.

Chapman was born in Massachusetts and graduated from Amherst College (1830). His degree was in the classics. He moved to Georgia and then to Florida in several tutoring or teaching positions. Sometime in the early 1840s he obtained medical training; of record is an honorary M.D. in 1846. He married Mary Ann Hancock, a widow, in 1839. They moved to Apalachicola, Florida, in 1847, which was home thereafter. She was a secessionist; he was a loyal Union person who helped fugitive slaves escape to the North. It does not sound like an ideal domestic situation.

Evidently the life of a successful physician was less strenuous than now for one often reads of eighteenth and nineteenth century botanists who stayed with their profession, medicine, but followed plants in their spare time. Chapman was one of these. He is said to have been an amiable, handsome, dignified man with unusually good health that continued into age beyond most of us. When eighty-three years young, he claimed to have walked thirteen miles to collect a rare orchid (was the thirteen miles round trip or one way?). He was a vivid descriptive writer and had a pictorial memory for details, a must for a floristic botanist.

The birth of Chapman's botanical interests apparently occurred while he lived in Georgia and was whetted by the then unknown wilds of north Florida. I presume he bought books and read them, but certainly he mostly obtained his plant training by doing. It is related that Chapman said to another physician-botanist, Dr. M.A. Curtis, that a southern flora equivalent to those for the northeastern states was badly needed. Curtis said yes, let's you and me do it. But Curtis didn't and Chapman did. Being professionally busy during the day, he led a double life, working on the manual at night. I suppose this is exaggerated just a bit. Surely he spoke to his wife now and then, about the price of peanuts if nothing else.

But it sounds slightly impossible. Knowing plants locally in Florida is a long way from possession of the necessary credentials for writing a flora of the southern states. How in the world could a man, isolated as he was, without a reference library, without a herbarium beyond his own gatherings, and apparently with but limited firsthand knowledge of the other states in the southeast (even though he had numerous correspondence contacts) possibly produce even a crude pioneer flora? But he did! He had manuscript in 1859. By his own account he went to Harvard and spent about five months there, consulting with Gray, making arrangements for publication, and then proofreading. Partly to the contrary, I find another reference that says Torrey proofread the work without Chapman's knowing. Which or maybe both? The fact that the opus (1860) was well done and filled a need is suggested by a second edition in 1884 and a third in 1897. The second edition has 698 pages in comparison to 621 of the first. Evidently quite a bit was added. In the preface to the second edition, Chapman stated that a third was planned in which he would catch up on nomenclature. But that was a bit too much. The preface to the third edition says, "The nomenclature of the first edition . . . is mainly retained." He has my sympathy.

No doubt Chapman's achievement can be viewed as little more than a pinprick in the story of plant science. But it held the scene for its region for some forty years. I am always in awe of those who do so much with so little. Chapman is remembered in the genus *Chapmania* (Leguminosae) endemic to his region.

SELECTED BIBLIOGRAPHY

Chapman, A.W. 1860. Flora of the Southern United States. Ivison, Phinney. 621 pp.
Ewan, J. 1971. A.W. Chapman. Dictionary of Scientific Biography. Vol. 3, pp. 196-197, with citations.
Mohr, C. 1899. Alvin Wentworth Chapman. Bot. Gaz. 27:473-478.
Trelease, W. 1899. Alvin Wentworth Chapman. Amer. Naturalist 33:643-646.

Charles Robert Darwin (1809–1882)

More has been written about Darwin than any other biologist who ever lived. The legions of books and papers concerning him include not only interminable analyses of his *Origin of Species* but also introspective biographies, the dynamics of the cultural and scientific climate in which he worked, his illnesses, work habits, and personal characteristics, what he thought of other people, and what they thought of him, etc.

Curiously, in light of this flood, he is rarely presented as a botanist. True, the fact that he wrote several books about his research on plants is mentioned in much Darwinia, but it is casual, somewhat in the light of "Well, the great man needs to play now and then." Indeed, I have been able to find only one paper (Ornduff 1984) that specifically treats his botanical work.

But Darwin was a botanist and in contact with botanists from the beginning of his career to the end; for example, Darwin wrote in his autobiography of the botanist **Henslow** at Cambridge that there was "a circumstance which influenced my whole career more than any other. This was my friendship with Professor Henslow." Later in life when he was putting together his meshwork of reasoning for the *Origin,* two of his most constant correspondents were **Asa Gray,** the American botanist, and **Sir Joseph Dalton Hooker,** premier plant geographer and systematist as well as director of the Royal Botanical Gardens at Kew. With them and others, Darwin winnowed the botanical evidence that went into his theories about

Charles Darwin

the mutability of species. Then more directly were some half dozen botanical books, mostly based on research the latter twenty years of his life, these alone placing him as the most important botanist of his time concerned with the reproductive biology of plants.

And that is the thesis here.

Darwin was born to reasonable wealth and married wealth, Emma Wedgewood, of the porcelainware family. Thus he had the advantage over most of us in having never to work for a living. Had he been employed at Iowa State University, not only would there never have been an *Origin,* but he would soon have been given walking papers because of zilch research production. Indeed he probably would not have survived graduate training. A facetious memorandum written by Professor R.W. Pohl to the dean of the Graduate College (Iowa State University, 1972) is cogent. I excerpt:

> One of our students, a Mr. Charles Darwin, is making unsatisfactory progress on his research. As you may recall, we sent him, with a full fellowship, on a several years cruise on the H.M.S. *Beagle.* He was supposed to prepare a dissertation on the marine fauna, to which he had excellent access during this voyage. Now, five years later, he has not only not completed his thesis, but fails to give me stipulated quarterly reports on his research. When this is discussed, he mumbles something about "needing time to be sure." He hints at an alleged new principle in biology, which he says is related to the discredited doctrines of a Professor Malthus. Although we have invested a considerable amount of time (and of the institution's money) in his program, I now feel that we no longer can continue to carry him along. He is of extremely retiring nature and is a wretched speaker. I do not think that he would be a satisfactory professor. At times he talks of entering the clergy, and perhaps that would be his best outlet.

After the *Beagle* trip, Darwin never traveled beyond the European continent, and indeed, but little beyond the vicinity of his home. There he lived quietly with his family the remainder of his life. His personality is well known from his correspondence and contemporary accounts. He possibly was the most modest, self-effacing, kindly biologist the world has ever known. That he has commonly been portrayed otherwise is evidence of cultural intolerance of views at odds with those socially prevalent.

Darwin, as an experimental botanist, worked primarily on pollination biology and its evolutionary implications. *The Fertilization of Orchids* was published in 1862, a second edition in 1877. Here

Darwin was the successor to the sequence of **Camerarius,** Kolreuter, and Sprengel, especially of the latter whose "wonderful book" he had read as early as 1841. And the implications of Darwin's study of orchids went far beyond that particular group of plants. Flowers not only constituted a mechanism for pollination by insects, as demonstrated by Sprengel, but most directly and usually for cross-pollination. The diverse adaptations of specific parts of the flower fascinated Darwin. "Nothing is useless," he said. "For each structure, there is a reason." This then lead to *The Effect of Cross and Self Fertilization* (1876). Here Darwin seems to have been the one to first expound the idea that later was called "hybrid vigor." Continuing, somewhat on the same topic, his *Forms of Flowers* (1877) dealt with heterostyly, unisexual flowers, and cleistogamy, the latter presenting the paradox of enforced self-pollination in contrast to obligatory cross-pollination. In all of these treatments, his data and thinking were landmarks for all future students of floral biology.

Then Darwin changed the subject to movement in plants, initially that of climbing plants. What factors are responsible for the behavior of tendrils as they seek an object to clasp and how does a pole bean know how to twine (*Climbing Plants,* 1875). Earlier **A.P. de Candolle** had demonstrated that stem movements were a result of exposure to light. If you blacked out the tips of the stem, it did not bend toward the light even though the rest of the plant was illuminated. Darwin extended Candolle's investigations and followed up with tactile response. Both workers presented the thesis that plant growth is mediated by the production of something in the growing tip, and that something moves elsewhere to do its work. They were perhaps pre-Avena coleoptile explorers. Darwin moved further in his *The Power of Movements in Plants* (1880) in which he studied various tropisms (reaction to light, gravity, etc.), and so-called sleep movements.

Yet continuing: insectivorous plants. The eternal experimenter worked with just one kind, *Drosera* (Sundew), and correctly deduced that stimulation of the tentacles was a response to the presence of nitrogenous compounds and the fact that "the plant should secrete, when properly excited, a fluid containing an acid and ferment closely analogous to the digestive fluid of an animal." Here apparently, he broke entirely new ground.

And finally, *Index Kewensis,* for which Darwin's money rather than his mind was instrumental. By the 1880s descriptive taxonomic literature had become so vast that no single worker could possibly keep up with it. Did the new plants in one's collection represent undescribed species or had they already been named by someone else? One could not readily know without an encyclopedic grasp of all literature—and access to it. Obviously impossible! Consequently, species and genera often were repeatedly described by different workers who lacked knowledge of each other's efforts. Plant nomenclature understandably was in turmoil. A bequest by Darwin made the initial publication of *Index Kewensis* possible. *Index Kewensis* was and is a citation index for the publication of names of genera and species of seed plants, 1753 to the present. The initial publication was in two mighty volumes (1893-1895); the manuscript was said to weigh more than a ton. (One doubts if it was quite that heavy.) It has been followed by sixteen supplements for subsequent intervals, and the availability of Number 17 is announced as this is written. Thus it continues as an essential standard nomenclatural reference for phanerogamous plants.

Perhaps all of this is small potatoes compared to the significance of Darwin's evolutionary theory. That is true. As a consequence, the weight of the *Origin* has obscured Darwin's botanical role, which is outlined here. Shorn of evolutionary theory, Darwin would be more clearly recalled as one of the significant pioneer contributors to botanical theory of the latter nineteenth century.

If one wants to read about Darwin, forget the current bull and try *The Autobiography of Charles Darwin and Selected Letters,* edited by his son Francis Darwin (1958). Most of it is letters; the autobiography, said to have been written for his children, is relatively short. Then, if yet interested, go from there.

SELECTED BIBLIOGRAPHY

Darwin, F., ed. 1958. The Autobiography of Charles Darwin and Selected Letters. Dover Publications, 365 pp.
    Republication of the book first published in 1892.
Eiseley, L.C.. 1956. Charles Darwin. Freeman Co., 10 pp. Reprinted from Scientific American, February 1956.
    A brief look at the total Darwin.
Ornduff, R. 1984. Darwin's botany. Taxon 33:39-47.

George Engelmann (1809–1884)

Not too many botanists beyond taxonomists know of Engelmann, and indeed information about him is sparse. But perhaps subordinate only to **Gray** and **Torrey,** he likely was the most substantial descriptive and monographic interpreter of North American plant wonders of the middle nineteenth century.

Engelmann, the senior of some thirteen children, was born in Frankfurt am Main, Germany. He finished an M.D. in 1831. European traveling brought him into contact with several name-botanists, including Karl Schimper and Alexander Braun, but he disclaimed their influence in bringing him to botany: "I began in my 15th year to be greatly interested in plants." He came to the United States in 1832, presumably to invest some money that had been entrusted to him by his uncle (Moore 1931). Instead it sounds like he spent money. He initially visited with **Nuttall** in Philadelphia, then to St. Louis from whence he took off on a lengthy botanical exploring trip about the adjacent states. (Another reference says the trip was to the southwest.) Then back to the St. Louis vicinity where Moore (1931) relates that he lived on a farm in nearby Illinois for a couple of years before returning to St. Louis and starting to earn money in medical practice.

Engelmann had now obviously decided to stay in the United

George Engelmann

States. He returned to the Old World only three times, the latter two visits primarily botanical, and the first (1840) seemingly was to get married. Anyway that is what happened; his wife was a cousin, Dorothea Horstmann, and they returned together to America. They had a son, George Julius Engelmann, who became an obstetrician and nationally known gynecologist.

Engelmann Sr. established a highly successful medical practice in St. Louis and initially probably devoted most of his time to it. In latter years, he was apparently sufficiently well off that he was able to cut back and devote most of his time to botany, including opportunity to travel to the yet unspoiled southwestern United States and see some of its wonders, particularly cacti, firsthand. Incidently he was also publishing in meteorology and was recording observations until the day before his death.

Engelmann's contacts with Asa Gray were initiated in an unusual manner. Engelmann sent duplicates of his collections to several German herbaria. Gray, who presumably had never heard of Engelmann before, saw some of these at Berlin. He wrote to Engelmann and the two were correspondents and, to some extent, collaborators in the years thereafter.

Engelmann was one of the founders of the St. Louis Academy of Science and was its president for some years. But much more importantly, he was instrumental in encouraging the wealthy St. Louis businessman, Henry Shaw, to develop his gardens about his country home (now far from the country!) for scientific use as well as public display. Gray wanted Engelmann to assume directorship of the garden, but Engelmann, saying that, "Shaw is as tough as any Scotchman . . . a man who has no real scientific zeal," declined. Not withstanding this pessimism, some of Engelmann's assertions must have rubbed off because the garden soon became an outpost of botanical culture in the yet primal Midwest. And Engelmann, so it is related, persuaded Shaw to call his establishment the Missouri Botanical Garden. So it is to botanists. But the synonym Shaw's Garden still persists, especially among the proud citizens of St. Louis. By either name, now more than 100 years later, the garden and its research facilities are among the most glamorous in the United States.

Engelmann is said to have been a slow and laboriously careful worker. His reputation rests on the fact that he was an excellent analytical taxonomist whose discoveries and judgments have well stood the test of time. He "did" *Cuscuta* (1842 and 1859). He worked up and published collections of Geyer (Illinois and Missouri), of Wislizenus (south Texas and Mexico), and, with Gray, those of Lindheimer (Texas). He is said to have published (1861) the first paper on plant pathology in the United States.

His greatest monographic works were in the Cactaceae—two major productions, which remain essential reference classics for cactus students to this day. He also did revisions in *Juncus, Quercus, Yucca, Vitus,* and *Pinus.* Studying *Isoetes* and various ferns he was contemporaneous with D.C. Eaton in establishing pteridology in the United States.

Dr. Engelmann's publications are sufficient, both in quality and quantity, to label him as one of our noteworthy botanical ancestors. Maybe we can also think of him as the stepfather of the Missouri Botanical Garden in its present form.

SELECTED BIBLIOGRAPHY

Anonymous. 1884. George Engelmann. Bull. Torrey Bot. Club 11:38-41.
Anonymous (possibly C.E. Bessey). 1884. George Engelmann. Science 3:405-408
Moore, G.T. 1931. G. Engelmann. In: Dictionary of American Biography. Vol. 6, pp. 159-160.
Shaw, E.A. 1986. Changing Botany in North America: The role of George Engelmann. Annals Missouri Botanical Garden 73:508-519.
  The essay here was written before this excellent reference was available.

Asa Gray (1810-1888)

Asa Gray dominated American botany some forty years. Furthermore he was not just an American botanist; by middle life he was an important world figure in floristic and revisionary botany, plant geography, and was a reasoned Darwinian exponent. Here we will first outline the generalities of his life, followed by his major taxonomic accomplishments, and lastly his role in the immediate post-Origin, Darwinian rubric.

Asa Gray, the first of three boys of Moses and Roxana Howard Gray, was born in Sauquoit, a village in upstate New York. His entire formal education was obtained within fifty miles of his birthplace, namely Sauquoit, Fairfield, and Clinton, all New York. While studying for his M.D. at the Fairfield College of Physicians, he encountered professors who excited him about natural history. Once the switch was turned on, the voltage never changed until he was immobilized a month before his death.

After receiving his degree, Gray only briefly practiced medicine. Instead, supporting himself by a succession of part-time teaching positions, he rapidly, even exponentially, expanded his knowledge of American plants. He was aggressive, soon making himself known to **Dr. John Torrey** of New York who was already somewhat of the father figure of the crescent American botany of the time. From a protégé, Gray rapidly emerged as Torrey's collaborator, then soon becoming the first American botanist who had stature equal to that of any of his European colleagues.

In 1842 Gray became Fisher Professor of Natural History at Harvard University, a position he held the rest of his life. There also he settled down by marrying (1848, age thirty-eight). One reads that his wife, Jane Loring Gray, was very delicate with numerous ailments as mysterious as those of **Darwin,** but also that she was the charming and vigorous entertainer of all and sundry who flocked to the Gray's doorstep in the ensuing years. There were no children.

Asa Gray was a small man, agile to his last years. Clean-shaven most of his life, he assumed the beard that one usually sees in pictures only in age. He had a perhaps dogmatic, somewhat fundamentalist religious morality and possibly a personal rigidity that Jane Gray is credited with softening. He was a magnetic personality to his professional peers, but evidently not so to students—there were none trained by him who subsequently rose to eminence.

Gray's life was frantically busy. This was a consequence of the combination of his personal fixation on a flora of North America and the myriad professional obligations coming to one of eminence, which he did not shirk. And from about 1860 to 1870, he was locked in mortal combat with the formidable Louis Agassiz concerning the Darwin matter.

Gray, age seventy-seven, was writing a scolding letter to **N.L. Britton,** that young upstart at Columbia University in December 1887, when stuck down, probably by a cerebral hemorrhage. He died the next month. Others rushed in but American botany was never the same again.

Now to Gray, the intrepid professional. He burst full-grown upon the American botanical scene as co-author with Torrey of the two initial volumes of a contemplated *Flora of North America* (1838-1843). Further publication was delayed because Gray's Harvard duties took him from New York, and because of the flood of new materials from the West, coming both to him and to Torrey. But a flora was in Gray's blood so he wrote one for the best-known part of the country. Thus came the *Manual of Botany of the Northern United States* in 1848. Gray put it through five editions. The currently used *Gray's Manual* of **Fernald** (1950) is the eighth edition.

However, that Gray had not given up a North American flora is evident not only in that he determined and recorded everything that came his way, but that he set himself up, somewhat dictatorially, as a clearing house for all American botany. This involved some management not only of collectors and other botanists, but also having his finger in the pie of all government-sponsored exploring expeditions. The resulting avalanche of critical publications, including descriptions of hundreds of new species, about North American botany has never been matched. These appeared mostly in the *Proceedings of the American Academy of Arts and Science* and U.S. government-sponsored exploration reports.

The mass of Gray's correspondence, U.S. and foreign, was awesome or awful, depending on viewpoint. The Gray Herbarium collection of letters addressed to its founder is said to exceed 10,000, and Gray presumably wrote at least an equal number. (There must be exaggeration here.) Although he kept no copies, many are preserved in various libraries across the country. And he had only a scratch pen, no access to an Apple Macintosh or IBM-compatible.

Gray established and maintained the herbarium, botanical library, greenhouse, and botanical garden mostly with his own hands. He received minor funding for the greenhouse and garden.

Because he was not a botanical evangelist for the unwashed, one almost wonders why Gray wrote botany textbooks. The addition to his income was not substantial. However, he dominated the textbook field for preparatory schools and colleges for perhaps forty years with titles such as *Elements of Botany, Introduction to Structural and Systematic Botany, How Plants Grow,* and others, several of these going through successive editions.

After his retirement from Harvard in 1873, Gray went back to the North American flora rainbow, with *Synoptical Flora of North America*. But his time ran out before its completion.

Gray's retirement and subsequent demise terminated an era. At Harvard, several individuals were employed to take his place. And with the change of the guard, those jealous of Gray's dominion over pelf and spirit (e.g., Greene and Britton) rapidly moved to claim the vacated turf. **Bessey** became the prime figure in textbook production, and German-trained physiologists a major force in the new American botany.

And now finally, Gray's contribution to biology in the broadest sense, one both eminent and effectual but largely unheralded. That is, Gray was a Darwin spokesman, fully equivalent of T.H. Huxley (Darwin's famed "bulldog") and perhaps even more significant. This came about through the convergence of discordant trails.

Darwin and Gray were avid correspondents and friends for years. Gathering botanical data relevant to his ponderings, Darwin asked Gray if he could furnish him with a comparison of the flora of Europe and that of the northeastern United States. Gray (1856) responded with a paper, *Statistics of the Flora of the Northern United States*. In 1859, he followed with a comparison of the flora of eastern temperate Asia and that of the northeastern United States. He found that some 40 genera of vascular plants were limited entirely to these two widely separate areas, and that, of 580 species in Japan, all had counterparts (the same or similar) in the northeastern United States. For explanation, Gray turned to historical geology. He postulated that during the Tertiary, a continuous temperate floristic unit ranged from Asia to North America, and that subsequent geological events had broken it into two segments (This thesis has stood the test of time.) Genera and species, Gray said, were not created separately in Asia and eastern North America. Those kinds, now represented in each of the two regions, were descended from one original stock, and their subsequent individuality was a consequence of change over time. Thus with Darwin, Gray was proclaiming the mutability of species.

The other factor was Louis Agassiz, Gray's zoological equivalent at Harvard and nemesis, a social lion and PR man who could raise $1,000 for his establishment for every $10 dollars raised by the reserved Gray. Agassiz, a pupil of Cuvier, believed the geological record to represent a series of independent creations followed by catastrophes that wiped the slate clean—all presumably planned and monitored by the creator. Agassiz's overshadowing reputation as the voice of science constituted a major threat to Gray's empirical and pragmatic philosophy.

The Darwin *Origin* was the breaking point. Agassiz ridiculed it as "very poor . . . very poor!" Gray, on the other hand, felt the Darwinian proposal sufficiently epochbreaking to abandon his reserve to directly challenge Goliath (i.e., Agassiz). He confronted

Asa Gray

Agassiz both in debate and independently at every opportunity, written or spoken. He bested Agassiz, not only by the merits of his case but because he marshalled his argumentation meticulously, while Agassiz, as always, depended on ad lib rhetoric and style. Gray followed with numerous, vigorous editorials in scientific and philosophical journals. These were published anonymously, but the content and structure betrayed the author. A clergyman, G. F. Wright, who saw in Gray a reasonable fusion of Darwin and Christianity, collected and published these in a single book, *Darwiniana* (1876). If it is true as Dupree (1959, p. 353) asserts that Agassiz "died a sick and bewildered man," it is one measure of the intensity of the contest.

Thus we see Gray not only the unique and dominant American descriptive botanist of his time, but as a world botanist, a philosopher of the natural sciences, possibly the first really important historical phytogeographer, and a foremost proponent of Darwin in the critical decade following the *Origin.* American botany, at least in these areas, now challenged that in the Old World.

SELECTED BIBLIOGRAPHY

Dupree, A.H. 1959. Asa Gray 1810-1888. Belnap Press of Harvard University. 505 pp.
  Excellent biography.
Dupree, A.H. 1972. Gray. In: Dictionary of Scientific Biography. Vol. 5, pp. 511-514.
  With citations; a synopsis from preceeding book.
Goodale, G.L. and S. Watson. 1888. A list of the writings of Dr. Asa Gray. Amer. J. Sci. Ser. 3. 36 (appendix):3-41.
Rodgers, A.D. III. 1944. American Botany 1873-1892, Decades of Transition. Princeton University Press. 340 pp.
  The center focus of the book throughout is Asa Gray.

Joseph Dalton Hooker (1817–1911)

The versatile Hooker, a combination of an explorer, a plant geographer, systematist, geologist, morphologist, and evolutionist, melded these diverse disciplines perhaps more effectively than anyone of his time.

The second offspring of **Sir W.J. Hooker,** he equally is one not to be forgotten. His childhood was spent in a household of botanical passion and possessions second to none other in England. Naturally, he also became a botanist? Of course not; he was trained as an M.D. (1839). But as elsewhere remarked, botanists are not made; they are born as such—even if their father was also a botanist. And in the breadth of achievements, J.D. Hooker considerably outmatched his progenitor.

Hooker married twice. His first wife, Frances Henslow, daughter of botanist **Henslow,** bore two children prior to her early death. The widower subsequently remarried and contributed to the production of two more Hookers. But this time the botanical line stopped.

Perhaps the most traveled botanical explorer of the nineteenth century, Hooker visited all continents, not just passing through, but usually working in yet unknown parts of the world for several years. He was assistant surgeon and botanist on his first trip, some four

Joseph Dalton Hooker

years, with the Ross expedition of the *Erebus* and *Terror* to the far reaches of the South Pacific and Antarctica. Only the voyage of the *Beagle* can exceed it in the fame of its discoveries. In midcareer, he became assistant director of the Royal Botanic Gardens, Kew, taking over many of the administrative duties as his father aged. He then succeeded him as director, 1865-1885. The remaining twenty-five years of his full life were devoted to more exploring, thinking, and writing. I wonder if the time span of his publications, 1837-1911, has ever been equaled among botanists, except perhaps by **L.H. Bailey.**

There are two volumes of published letters of Joseph Hooker (Huxley 1918). These letters, particularly from his middle years, are the most engaging, among botanists, I have ever read. Hooker was a forceful and colorful writer with an epigrammatic gift for words and the ability to turn a neat phrase. He wrote of everything. I find him thus more fascinating than when presented secondhand in the biographical *Hookers of Kew* (Allen 1967). He is portrayed by one writer as being proud and autocratic. Perhaps so. Certainly Hooker knew his value, and he assuredly could cut an austere visage when necessary. But he was also as kindly and broadly philosophical an individual as one could encounter—read some of his letters.

A frugal account cannot do justice to Hooker's long career. Of the report of the Ross expedition, he produced three botanical volumes about the circumboreal flora of the South Pacific and how it got that way. They established him as a major world botanist. Several years in India and Nepal was the stimulus for the Hooker-edited, seven-volume *Flora of British India* (he was author in part as well as editor). His *Himalayan Journals* are an engaging popular report. Then he finished Trimen's *Flora of Ceylon* after the author's death. And then came the monumental **Bentham**/Hooker *Genera Plantarum*. Although discussed in an essay on Bentham, I partly repeat here. Issued in several sections that were assembled into three volumes, 1882-1883, the work treats 7,565 genera of flowering plants in 6,617 pages. A fair job! Although now much dated (and written in Latin), the work remains a major reference that has yet to be replaced.

Hooker was one of **Darwin's** few intimate friends. Darwin

shared his natural selection thesis with Hooker some years before publicly revealing his brand of heresy. From his own work, Hooker subsequently confirmed Darwin's ideas in the process of explaining the distribution of the same or seemingly closely related plants across the islands of the South Pacific. Knowing nothing of modern plate tectonics, he hypothesized land bridges but later inclined towards Darwin's chance long-distance distribution. And along with Thomas Huxley and **Asa Gray,** he was one of the triumvirate of the most important Darwin supporters in the immediate post-origins imbroglio.

As an administrator, J.D. Hooker followed his father in rendering Kew a mecca for botanists. For example, he broadened its scope in persuading a wealthy friend, T.J. Phillips Jodrell, to fund the establishment of a laboratory for morphology and physiology.

Rubber production was then a Brazilian monopoly. To some extent, Hooker was the figure behind an Ollie North-type abduction of rubber tree seedlings from Brazil. These, propagated at Kew, then went to the British colonies in tropical Asia and initiated the establishment of efficient and affluent rubber plantations in that part of the world.

Somewhere along here, as was his father, Hooker was knighted and became Sir Joseph Dalton Hooker.

I mentioned the origin of *Index Kewensis* in my account of Darwin, who largely financed the original edition. This was and is a citation index for the place of publication of all names of genera and species of seed plants—1,753 to the time published. The mighty volumes came out 1893-1895. Hooker, now an elderly man, is said to have personally read proof on the entire compilation. This consists of about 2,500 pages, each page with three columns, each with about fifty names. This is hard to credit. But both Hooker and Darwin live on in this monstrous production, one essentially anonymous except to systematic botanists. The original has been followed by 17 supplements.

No biologist in the nineteenth century won either the public praise or the eternal damnation that came to Darwin. But Hooker matches Darwin in being representative of the best in British botany 100 years ago.

Joseph Dalton Hooker

SELECTED BIBLIOGRAPHY

Allen, M. 1967. The Hookers of Kew. 1785–1911. London. 273 pp.
Bower, F.O. 1913. Sir Joseph Dalton Hooker. In: Oliver, F.W. ed., Makers of British Botany, pp. 301–323. Cambridge University Press, London.
Desmond, R. 1972. J. D. Hooker. In: Dictionary of Scientific Biography. Vol. 6, pp. 488–492.
Huxley, L. ed. 1918. Life and letters of Sir Joseph Dalton Hooker. Arno Press, London. Two volumes.

Carl Wilhelm von Nägeli (1817-1891)

Nägeli, son of a physician, started to follow in his father's footsteps but soon switched to botany. He also spent a year studying the philosophers, particularly Hegel, whose influence, it is alleged, permeates much of his writing. He successively had positions at Jena, Zurich, Frieburg, and finally Munich. Although I have plodded through several references, I find nothing beyond his intellectual achievements save that he married in 1845 and spent his honeymoon collecting algae.

Nägeli left life, so we are told in an obituary notice, "the most illustrious of that illustrious band of botanists, to whom our chief advances and science are due" (Scott 1891).

But fame may be temporal. After a century now, Nägeli is scarcely known except as the great botanist to whom the humble **Mendel** presented his work and who, in effect, told Mendel to go chase himself. Asimov (1971) has damned him, saying "Nägeli did far more harm to biology than good," based not only on his Mendel blindness but his visions of orthogenesis and teleology. I moderate Asimov's violence in suggesting that probably no one else would then successfully have "read" Mendel; the theoretical botanists of the middle eighteenth century were mostly holistic—rather than reductionist-minded. But later, among the three "discovers" of Mendel was Nägeli's pupil, Correns.

Carl von Nägeli

Nägeli published voluminously on nearly all of the hot issues of his time with observations and new ideas. The observations were mostly excellent (he was a skilled microscopist), the ideas not necessarily so. But this alone gives a likely reason why he is so little known today. Individuals like **Darwin** and Mendel remain in the headlines because of a single overpowering issue. Nägeli's genius (if it was such) was diluted not only by the diversity of his efforts, but among the work of others, in the incremental conceptual growth of issues relating to our understanding of the mechanisms of plant life. In the latter, Nägeli was but one in a flow chain of understanding (sometimes also unfortunately of misunderstanding).

For example, the cell problem: **Schleiden,** whose name is usually associated with cell theory, however, left it messy. He had propounded that cells produce the new by internal budding. **Von Mohl,** concentrating on cell wall formation, showed that this was not so. Nägeli went after the nucleus. Using improved staining, he found it an almost invariable component of plant cells. He observed that it divided concomitant with cell wall formation; the wall was the result of division, not the cause. He undoubtedly saw chromosomes, but an understanding of mitosis had to await further advances in microscopy and staining techniques.

Nägeli puzzled about sexuality in all groups of plants. He observed antheridia on fern prothalli and saw ciliate sperms and other motile cells in various cryptogams. This confused him. If the leafy mosses produced sex organs, why didn't the leafy ferns also do so; the sex organs were not where they belonged. Nägeli assembled much underbrush but never saw the pathway through it. It took **Hofmeister** to do that.

But Nägeli developed the distinction between meristematic tissue (Bildungsgewebe) and structural parts (Dauergewebe) whose cells do not multiply. He divided tissues into trophoplasm (not hereditary) and idioplasm (hereditary material). He also wrote on the causes of evolution in which he envisioned an orthogenetic (vitalistic?) striving for perfection; natural selection indeed trimmed off the misfits but that was viewed as secondary. He believed in spontaneous generation, apparently even post-Pasteur.

Nägeli studied apical growth in vascular plants. He found orderliness in cell derivation lineage from the apical cell. This was the kind of a thing he was looking for, underlying laws that could be expressed in mathematics. He presented a complex "micellar" theory for cellulose cell walls (and for starch grains). Although molecular structure was envisaged and given (Nägeli called it atomic), his thesis was based primarily on gross structure and swelling properties. With some modifications, his formulation has been widely accepted as the intussuception thesis of cell wall formation. Sweet theory was here dominant; neither Nägeli or anyone else had actually seen a starch or cellulose molecule.

As previously stated the undercurrent of teleology and vitalism that exists throughout Nägeli's work presumptively derives from Hegel. Nägeli, however, has explicitly denied this.

The above is sufficient to portray Nägeli as a potpie mixture. Olby (1974) says that "he deserves sympathetic evaluation both as an innovator and a victim of the biological thinking of his time." True perhaps also of botanists of the present.

SELECTED BIBLIOGRAPHY

Asimov, I. 1971. Asimov's Biographical Encyclopedia of Science and Technology, 2d ed. Doubleday and Company, New York.

Morton, A.G. Nägeli. In: History of Botanical Science, pp. 384-387. Academic Press, London .

Olby, R. 1974. C. Nägeli. In: Dictionary Scientific Biography. Vol. 9, pp. 600–602, with citations.

Scott, D.H. 1891. Nägeli. Nature 44:580–583.

Johann Gregor Mendel (1822–1884)

The geneticists reasonably claim Mendel. But his work was made possible only because he was a good botanist. I presume everyone in biology knows who Mendel was, and that he did genetics before there was any genetics. I also assume (though with reservations) that you also know what he did and what he proved. Therefore, I pass this by and talk of other things.

Johann Mendel (he took up the Gregor later) was born of peasant parents in Silesia. His farm background gave him familiarity with both animals and plants at the everyday work level. His secondary education was fragmentary and interrupted. As there was little money for sending him off to high school, he supported himself in part by private teaching. Then he gave up. He entered an Augustinian monastery at Brünn (now in the Czech Republic) as a way of completing his education without also having to struggle to keep body and soul together. After a few years he took examinations for accreditation as a teacher. He failed twice. So he taught natural science on an unaccredited basis until he became the abbot (or prelate) of the monastery. To all accounts he was a good teacher. It is related that he would take a pocket of peas with him to classes and hurl them at students who started to doze.

Mendel's interest in inheritance in plants and animals derived from his boyhood days. He started some animal breeding experiments. The superior bishop put a stop to this on the basis that he was improperly tinkering with sex. So he changed to peas. Mendel is reputed to have said, "I turned from animal breeding to plant breeding. You see, the bishop did not understand that plants also have sex." This sample of humor perhaps supports the characterization of Eakin (1975) as a "jolly priest . . . with a taste for secular pleasures, two of which were food and cigars." But I fear Eakin's characterization of Mendel as a jolly priest is off base. Other references (Kruta and Orel 1974) indicate that Mendel was subject to continued bouts of depression (the probable reason he did not pass his teaching exams) and that likely he was not especially outgoing in personality. Neither have I seen any further mention of his secular interests or pleasures.

Mendel became rotund in his fifties and could no longer easily bend down to tend his plants. Possibly this was an ancillary reason for giving up his garden pleasures.

After preliminary work, Mendel began eight years of study of the inheritance of selected qualitative characters in peas. He published the results in 1866 in the Brünn *Society for the Study of Natural Science* (translated). Recently it has been stated (Stern and Sherwood 1966) that the Mendel paper "is one of the triumphs of the human mind . . . it presents facts in a conceptual scheme which gives them general meaning . . . it is a supreme example of scientific experimentation and profound penetration of data." However, a dog must bark on the right day or in the right place to be heard, because in 1866 the paper caused less excitement than a local rainstorm. Mendel did attempt to bring attention to his work through correspondence with the great botanist **Nägeli.** It was futile. Nägeli encouraged him to work on *Hieracium* (hawkweed), a relative of the dandelion, with the implication that there he might learn something of significance. Mendel barely got started on this—just as well, *Hieracium* is apomictic (produces seed without fertilization). He quit; probably because of discouragement and because he had now advanced to administration, prelate of the monastery. Local people vaguely remembered him otherwise as a pleasant eccentric who liked to grow peas just for fun. Incidently Mendel was also interested in meteorology and published several papers in that field. One assumes they were not as theory-encompassing as his genetics later

Johann Gregor Mendel

turned out to be.

Mendel took his duties as prelate seriously. For example, a new law subjected monasteries to a tax that Mendel considered illegal. Other monasteries caved in but Mendel fought it to the end. And he expanded the service of the monastery to the surrounding peasant population. For these reasons and others he was regarded as a kindly and effective abbot not only by those of his order but by the clergy of other faiths of his region. And he was popular with the peasantry of his area. As I said, they knew that he had played with peas, but this was regarded tolerantly as one might look upon a busy person who none-the-less takes time to collect bottle tops.

And so Mendel departed this world in 1884. During the rest of the latter years of the nineteenth century, his publication was casually cited a couple of times but without any substantive interpretation.

Most know that Mendel was "discovered" almost simultaneously at the turn of the century by von Tschermak, **de Vries,** and Correns. If he had not been discovered, someone (perhaps one of those named) probably would have duplicated his work. The time had come; the cytological bases for Mendel's findings, i.e., chromosomes and reduction division, were now known and fitted into Mendel's laws, hand in glove. This catenation would follow, however, only if the subject characters were on different chromosomes. And Mendel by blind chance, lottery luck, hit it. Was it somehow predestined? In any case, his work was that which, by subsequent saltations, ignited the flames that became the science of genetics.

So now we had both genetics and Darwinian evolution. But they were not necessarily good bedmates. The capability of Mendelian genetics to produce the kind of variation that would be effective in the "struggle for existence" was not at all clear. Was genetics really **Darwin's** missing link and did it support evolutionary theory? Answering that question, to the accompaniment of much controversy, took another forty years. It was only with the advent of the so-called Neo-Darwinism that the evolutionary geneticists and the Darwinists were able to arrange their cards in a single consonant hand. The development of population genetics and population biology the last twenty years has further rendered these once independent, even combative, disciplines one of unitary concept.

Over these years Mendel's stature grew from nothingness to that of a demigod: an amateur, who in his spare time accomplished one of the greatest feats in biology before anyone else even had an inkling of what it was all about! But the role of a god is a precarious one and brings debunkers be there any opportunity. The famed statistician, R.A. Fisher, in 1936 published a paper stating that a statistical analysis of Mendel's data suggested that the fit to theoretical predictions was just too good to be due to chance—in other words, presumably Mendel had fudged his figures when they wandered a bit beyond his expectations. Fisher's paper resulted in a minor spate of literature that has been briefly analyzed by Orel (1968). Orel's conclusion was that "There are no proofs either of the possibility that Mendel had been deceived by somebody, or that he had unconsciously or deliberately falsified the data from his experiments." Orel was vigorously supported by Dobzhansky who implied that Fisher had somewhat jumped the gun. I agree. We can only acknowledge that Mendel did have some luck in his unknowing choice of characters.

Mendel, who couldn't even pass a teaching examination, evidently could never have gotten into graduate school. Had he lived now, as a release from his serious duties, he probably would have watched TV, perhaps becoming a Chicago Cubs fan. But without such distractions, he has had few peers in the use of an analytical methodology from which he built a viable theory.

SELECTED BIBLIOGRAPHY

Boyes, B.C. 1966. The impact of Mendel. Bioscience 16:85-92, with citations.
Eakin, R.M. 1975. Great Scientists Speak Again. University of California Press, Berkeley, 119 pp.
Kruta, V. and V. Orel. 1974. G. Mendel. In: Dictionary of Scientific Biography. Vol. 9, pp. 277-283, with citations.
Orel, V. 1968. Will the story "too good" results of Mendel's data continue? Bioscience 18:776-778.
Stern, C. and E.R. Sherwood, eds. 1966. The Origin of Genetics. A Mendel Source Book. W.H. Freeman, San Francisco, Calif. 179 pp.

Nathanael Pringsheim (1823–1894)

Pringsheim was one of the influential voices among his generation of the combative Germans who pioneered in expanding the realms of botany beyond systematic theory and description. Although his most important achievement was probably that of extending **Hofmeister's** alternation of generations to the algae and fungi, he also contributed to plant physiology and cellular theory.

Pringsheim got into botany through stubbornness. His father was a successful business administrator and wished the son to be likewise. Junior was not enthusiastic. So papa proposed that if he must study science, he best go into medicine where there were good chances of making an excellent living. The younger Pringsheim gave it a try. But following a less than exciting start in medical school, he switched to botany getting a Ph.D. (1848) rather than an M.D.

Pringsheim married in 1851, the "happy union" producing three daughters and lasting forty years (he outlived his wife by a year).

Now with the doctorate union card, Pringsheim moved from an appointment as Privatdozent at Berlin to that of a professor at Jena but this did not last; evidently he did not like teaching, did not get along with administrators, and had episodes of poor health.

When in 1868 his father died, he came into a considerable inheritance. Unlike others, he now was no longer dependent on a salary. He quit! He moved to Berlin, where he equipped his home with a private laboratory in which he continued his life work.

Despite this seeming isolation as now an amateur botanist, Pringsheim was influential in interaction with the system. For example, he founded the journal *Jahrbücher für Wissenshaftlich Botanik* (well known to anyone who has entered the German literature) and served as its editor to the end of his life. He participated in the creation of the German Botanical Society and became its president.

Pringsheim's research seems to fall into three periods. The first, picking up from his doctorate thesis that dealt with cell wall formation (apposition from the interior, not from the outside), concerned the continuing saga of cell theory. Successive workers had seen mostly what they expected to see and asserted that every one before them was an idiot. Following this formula both Pringsheim and **von Mohl** undertook to correct the cell multiplication "watchcrystal" thesis of **Schleiden.** But the two reformers did not agree with each other, and Pringsheim next took von Mohl to task for his obtuseness. The curtain on this ado about something came only in the latter quarter of the nineteenth century when the details of mitosis were finally worked out.

Then Pringsheim moved to life history studies, especially of the algae that Hofmeister had barely touched. For some years he worked with a wide variety of both marine and freshwater kinds, as well as the fungus *Saprolegnia*—on the latter his observations collided with those of **de Bary.** He identified motile "spores" as sex cells. The terms oogonia and sporangia come from him. He observed the penetration of the egg by the sperm in several genera, reasonably establishing that there was an organic fusion, not just a stimulatory effect provided by the sperm.

To link these findings with the higher cryptogams, he studied the moss protonema and the life history of *Salvinia,* a fern. Leaving observation for theory, he subsequently concluded that an alternation of life forms, or its equivalent, is evident in diverse algae, and thus is nearly universal except for the blue-green algae. He then proposed the so-called homologous theory of alternation of gener-

Nathanael Pringsheim

ations as a means of linking the algae with the bryophytes and providing unitary concept for the entire plant kingdom. His observations were mostly confirmed by others, but the homologous concept was speculative. It was immediately challenged by an "antithetic" theory of Celakovsky that asserted that the bryophyte sporophyte is not homologous with the fusion-derivative phase of the algae—rather it was *de novo*. There was now considerable noise about these two postulates. **Bower** (from England) later supported Celakovsky on the basis of further observations, and the original controversy seems since to have faded away from fatigue except for its remembrance in plant morphology courses.

In the latter part of his life, Pringsheim moved to the function of chlorophyll. Again he sparked contrary views because he came to the conclusion that chlorophyll had little directly to do with photosynthesis. A reasonable understanding of the role of the chlorophylls really was not achieved until the time of **Willstätter** and Stoll early in the present century.

This was the time in which the German students of life histories and physiology, despite numerous goofs, vastly expanded the scope of theoretical botany. Pringsheim was a major actor.

SELECTED BIBLIOGRAPHY

Geison, G.L. 1975. N. Pringsheim. In: Dictionary of Scientific Biography. Vol. 11, pp. 151-155, with citations.
Sachs, J. von. 1875. Geschichte der Botanik, Munich.
    English Translation, Garnsey, H.E.F. and J.B. Balfour. 1890. History of Botany, p. 567.
    Index leads to Pringsheim.
Scott, D.H. 1895. Nathanael Pringsheim. Nature 51:399-402.

Wilhelm Friedrich Benedikt Hofmeister (1824–1877)

Most biological concepts have evolved over time, sequentially passing from one worker to another. Those of the cell and of evolutionary mechanisms are good examples. But that of plant life cycles, alternation of generations, burst essentially full grown from the eyes and mind of Hofmeister, one of the more remarkable botanists of all time.

Hofmeister's father, a music publisher in Leipzig, was an amateur botanist. The young Hofmeister's formal education terminated with high school and he went into his father's business; indeed, he continued as a commercial publisher to the end of his life. Hofmeister married in 1847; there were nine children of whom six grew to maturity. His wife died in the early 1870s; he married again less than a year before his death. He is said to have been "short, dark and extremely vivacious" (Proskauer 1972). He was shortsighted but refused to wear glasses (did this myopia help him in the kind of observations he made?).

The origin and nature of Hofmeister's botanical interests, possibly from his omniverous reading are not clear to me. Perhaps we can credit **Robert Brown** whom Hofmeister held in high esteem. Beyond the fact that he developed into a skilled preparator of material and microscopist, there must have been some keen idea or basis for the use of these skills.

Wilhelm Hofmeister

Seemingly Hofmeister was a lone amateur who worked primarily early in the morning (4 to 6 A.M.) before the routine day began. Following preliminary publications, his work was published as a book with a long title that one boils down to *Vergleichende Untersuchungen* (Comparative Researches). It appeared in 1851, while he yet had no professional affiliation.

What then was this *Comparative Researches?* Hofmeister studied embryo sac and pollen tube development in some 40 Angiosperm species representing 19 flowering plant families. There he clearly demonstrated (both by words and illustrations) that the embryo develops from the fertilized egg cell, not the pollen tube as **Schleiden** had said. Then he started comparing sexual reproduction as seen in the flowering plants with that in ferns, *Equisetum*, *Lycopodium*, mosses, and liverworts. Hofmeister found the archegonia and antheridia (the sex organs) in all of these diverse kinds of plants. From this and observation of what grew from them, he concluded that there is an alternation of two kinds of plants in the life cycle. The stalked sporangium of the moss is the equivalent of the leafy fern plant that, like the moss sporangium, produces spores without sexual fusion. The liverwort thallus is the equivalent of the gametophyte of the ferns, producing both of the sex organs, the antheridia and archegonia. The egg and sperm then, after fusion, grow into the other kind of a plant, the sporophyte. Among heterosporous kinds, like *Selaginella*, the gametophyte develops primarily within the spore. And here is the stepping stone to the seed plants, the Gymnosperms and Angiosperms, also heterosporous, the difference being that in these latter groups the spore is never shed from the sporangium. The female gametophyte comes to be called the embryo sac.

Others had floundered around with some of these observations, but none had even approached putting it in a unified structure. All of this then by one man. And it was all the more remarkable because Hofmeister's work was before the development of sophisticated microtechnique, stained slide preparation, the use of the microtome, etc., that came the last half of the nineteenth century.

The preceding demonstrates a superb combination of both observation and its interpretations for a unified concept for the plant kingdom. Almost, Hofmeister could walk to a lecture podium today and present alternation of generations with but footnotes of qualification. All of this puts his other botanical accomplishments somewhat into shade. But turn the light on and they are there. In fact his first work had to do with the demolition of Schleiden's thesis of cell multiplication through budding, and the idea that the embryo is male—derived from the pollen tube. Hofmeister also studied cell structure, the movement of sap (water with dissolved solutes) in plants, and other physiological entrées.

Naturally this drew attention. Hofmeister was abruptly a major botanical figure. He was awarded various honorary degrees and ascended first to the botanical chair at Heidelberg in 1863, moving to Tubingen following **von Mohl's** death in 1872.

Now an esteemed professor, he should have been able to fly uninhibited to yet greater heights. But sadly it was anticlimax. He started a handbook series on physiological botany, but it was generally viewed as second rate. Perhaps his latter major contribution to posterity was his pupil, **Goebel.** He was still in the publishing business and his health was declining. Following strokes he died in 1877, young compared to the life span of many botanists.

SELECTED BIBLIOGRAPHY

Larson, A.H. 1930. Wilhelm Hofmeister. Plant Phys. 5:612-616.
Morton, A.G. 1981. W. Hofmeister. In: History of Botanical Science, pp. 398-404. Academic Press, London.
Proskauer, J. 1972. W. Hofmeister. In: Dictionary of Scientific Biography. Vol. 6, pp. 464-468.

Heinrich Anton de Bary (1831-1888)

To the extent that anyone can be said to be the founder of anything, de Bary was that for plant pathology. But no, says Whetzel (1918), he was a botanist, not a pathologist. Some say instead mycology, for example, Robinson (1981): "He was the founder of that branch of botany." But there he was preceded by **Persoon** and **Fries.** However this may be, he elucidated the life histories, comparative morphology, and relationships of the fungi on a comprehensive basis, far beyond anyone else of his time. Mycology with him, indeed, was established on a firm basis.

De Bary, German, was one of ten children. His degree was in medicine but simultaneous with preparation of his dissertation, he also produced a book on the rusts and smuts (1853). He married in 1861; there were four children. Proceeding through a succession of appointments, he ultimately became professor at Strassburg (ca. 1871) where he spent the remainder of his life.

The first major contribution that de Bary made to mycology was to yank the fungi out of the never-never world of spontaneous generation. The idea yet existed that given suitable conditions, microorganisms, molds, worms, and even mice would spring out of nothing to become what they were. A firm subscriber to Pasteur, de Bary demonstrated that fungus growth arises from a preexisting spore or similar body. Also and contrary to popular belief, he showed that fungi were the causes, rather than the spontaneous results, of plant diseases.

An important example of de Bary's work is his extended studies of the scourge of potato blight and its causative agent, *Phyophthora*. He traced the entire life history of the organism, ultimately using spores to infect healthy plants, the Koch's postulate procedure. Then he attacked the more complex problem of black stem rust of wheat. Perhaps as a consequence of observing the frequent propinquity of the barberry to areas of infestation, he demonstrated that the *Puccinia* needed both the barberry and the wheat for the completion of its life cycle, and that infection of wheat derived from the barberry. He termed fungi like *Puccinia* heteroecious.

Otherwise, de Bary showed that lichens consist of an alga and fungus living together for mutual happiness and designated this lifestyle as symbiosis. He reached throughout the fungi, even to the then imponderable myxomycetes, slime molds. George Martin (University of Iowa, two generations ago; unmatched Myxomycete specialist) has written admiringly of de Bary's studies of the slime molds. According to Martin (1958), de Bary's findings were especially spectacular considering the methodological limitations of his day; de Bary carefully distinguished between hypothesis and verified fact, and his publications remain classics of enduring value. Martin's commentary, though specifically about the Myxomycetes, can be seen as reasonably applicable to the quality of de Bary's work generally.

De Bary went beyond the fungi. He wrote a book on bacteria, summarizing knowledge to date (1885). He and his students studied the morphology of vascular plants (*Comparative Anatomy of Ferns and Phanerogams*) wherein Morton (1981) credits him with the organization of the "principal features of mature plant anatomy on a secure foundation."

Like Beethoven, de Bary died at age fifty-seven. He enters no history (as will Ronald Reagan and Saddam Hussein) save of the most specialized kind. But which of these three has contributed the most to human welfare and knowledge?

Heinrich de Bary

SELECTED BIBLIOGRAPHY

Anonymous. 1988. Anton de Bary. Nature 37:297-299.
Martin, G.W. 1958. The contribution of de Bary to our knowledge of myxomycetes. Proc. Iowa Acad. Sci. 65:122-127.
Morton, A.G. 1981. History of Botanical Science, p. 429. Academic Press, N.Y.
Robinson, G. 1981. Anton de Bary. In: Dictionary of Scientific Biography. Vol. 1, pp. 611-614, with citations.
Whetzel, H.H. 1918. An Outline of the History of Phytopathology, pp. 45-47. W.B. Saunders, Philadelphia, Pa.

Julius von Sachs (1832–1897)

More than any other, Julius Sachs was the forceful man who grabbed the pieces of plant physiology and put them together as a discipline—one that soon was in competition with the then dominant systematics. He was more than a physiologist; he was the epitome of the "New Botany" and his *Lehrbuch der Botanik* was the bible of its time.

Sachs was German. His father was a woodcut engraver. Both of his parents died when he was yet in high school (gymnasium) and his education consequently was interrupted. Struggling to make a meager living, he had the good fortune to work for the famous animal physiologist Purkinje. Presumably this provided both the opportunity to finish his education and find his life's direction. Once on the road, he read philosophy, turned out research papers like an animated machine, and was starting to write books. After a succession of appointments he accepted the position at Würzburg in 1868 where he stayed, refusing several more prestigious offers.

One would have the idea that his personal life was minimal. All I find is that he married "a lady from Prague" and that he had a daughter, Maria, who was an artist. Sachs himself was a brilliant, opinionated, and egocentric fireball. In his later years he became harsh and implacable and would not tolerate opposition to any of his theories. He was a dazzling, albeit caustic, lecturer. He accumu-

Julius von Sachs

lated diverse health problems (as come to many of us), and one correspondent stated that letters from him in the last fifteen years of life were "one protracted health report."

Precisely what did Sachs do to achieve the levitation of both plant physiology and self? A recitation sounds unconvincing because it cannot elucidate the nature and impact of his work. Initially his efforts were devoted primarily to opening new vistas in plant physiology wherein his originality, experimental precision, and powers of synthesis outpaced any of his contemporaries. For example, he described (within the limits of the chemistry of his day) the metabolism of seed germination. He defined the essential mineral nutrients of plants and grew them from nutrient cultures. His studies showed that the chloroplast is the site of photosynthate production of which chloroplast starch is the first visible product; starch accumulates during the day in the presence of light and CO_2, and then disappears at night. He distinguished between substances that could be recycled (fats, proteins, starch), and those not biodegradable within the plant (e.g. cellulose, lignin, wax). He was concerned with the interrelationship of the plant and the physical aspects of its environment, for example, temperature and light. And he participated in other battle fronts, for example, the homologous vs. antithetic theses of alternation of generations, rife in his day. The combination of his skill as a microscopist and artist, his keen knowledge of the chemistry of the time plus precision of experimentation—all of this placed him in the front rank of all botanists.

But this research was only the beginning. Sachs also wrote books that were catalytic in the emergence of twentieth-century botany. Perhaps the most important was his *Lehrbuch der Botanik* (1868), which went through multiple editions and translations. It quickly became divine creed for the botanical progressives. For example, **Bessey** in the United States, soon a Sachsophilian evangelist, interjected the philosophy of the master into numerous editorials and subsequently into his own textbook writing.

The *Handbuch der Physiologischen Botanik* (1865) and the *Vorlesungen Über Pflanzenphysiologie* (1882) were Sachs' statements about his special realm. His *Geschichte der Botanik (History of Botany)* was also translated into English. There his views of many of his predecessors are unkindly, but kindliness was not always one of Sachs' attributes. An effective philosopher of ideas, he had numerous students who became apostles of the faith. Indeed, in physiology, both in England and subsequently in the United States, you were not with it unless part of your training was in Germany and especially with Sachs.

Along with this paean of praise for accomplishment, one must allow that Sachs did occasionally strike out, this perhaps mostly in his latter years when he considered himself sufficiently an authority that he could perceive the truth without testing. A prime example is his writing on the ascent of sap, i.e., how water moves up to the top of a tall tree. In a paper on the porosity of wood (1879), he said, "Water movement depends . . . upon imbibition and swelling . . . upon the motion of the . . . water molecules which are contained between the micellae of the wood cell walls. This can only occur when the wood cell walls at the upper end of the system lose a portion of their water molecules . . . attracting water from the nearest wood cells . . . this movement, extending backwards . . . proceeds from the foliage of a land plant to the roots which absorb the water from the earth." This thesis, based seemingly on overtones of osmosis, was apparently conceived independent of any experimental evidence. Some laboratory work easily could have proved that the author, rather than the cell walls, was all wet. Others skeptical of the Sachs pronouncement quickly tested it experimentally by blocking the tracheal lumina with substances such as cocoa butter or paraffin. The leaves of the subject test plants wilted.

Initially, Sachs reacted favorably to **Darwin** but soon became devestatingly critical. This deplorable hostility to another great genius of this time as been reviewed by **Tansley** (1934), who devoted several pages to a subfreudian analysis of the manner of Sachs' "habit of mind," personality, and research methodology. Whatever the rationale, however, one cannot avoid expecting better from a scholar of Sachs' stature. Sadly the more honors were heaped on Sachs in his latter years, the greater grew the cancer of his bitterness to the universe about him.

But a prize hog is yet a ribbon winner, warts and all. Sachs might have said that he broke the tyranny of taxonomy. I will say

Julius von Sachs

that probably, in his time, he did more to shape the future course of botany than any other single individual.

SELECTED BIBLIOGRAPHY

Bopp, M. 1975. J. Sachs. In: Dictionary of Scientific Biography. Vol. 12, pp. 58–60, with citations.

Pringsheim, E.G. 1932. Julius Sachs, der Begründer der Neuen Pflanzenphysiology, 1832–1897. Gustav Fischer Jena. 302 pp.

Tansley, A.G. 1934. The founder of modern plant physiology. Phytologia 33:232–240.
Purportedly a review of Pringsheim's biography of Sachs, but mostly an independent essay.

Weiss, F.E. and collaborators. 1933. Commemoration of the centenary of the birth of Julius von Sachs. Proc. Linn. Soc. London 145:1–7.
Eulogies of Sachs include essays by Vines, Scott, and Bower.

Julius Oscar Brefeld (1839-1925)

Brefeld, German, seemingly little known at present, was one who devised pure culture methods used in mycology. Immediately following **de Bary,** he became the most important student of the life histories and comparative morphology of the fungi.

A scan of Brefeld's personal life is a sequence of disasters. Initially trained as a pharmacist, he obtained a Ph.D. in botany (1868) and briefly studied with de Bary. Then, debilitated by pneumonia, he was a semi-invalid for a couple of years during which he studied art in Italy. He turned back to fungi but was now interrupted by the Franco-Prussian War during which he contracted typhoid fever. Then a retina detachment and glaucoma resulted in the removal of one eye. Again there was a break in Italy with art. He married at the age of fifty-seven; his wife died six years later. Glaucoma in his remaining eye sufficiently interfered with his teaching duties (professor at Breslau) that it was necessary for him to resign (1907). He was completely blind by 1910 but continued uncompleted publication by dictation.

Brefeld's formal positions included a sequence of Privatdozents and professorships, that at Munster (1884-1902) being of longest duration. He is said to have been less than diplo-

Julius Brefeld

matic, possessed of acidic wit, polemic both in conversation and writing, and unkind to those who disagreed with him. Perhaps the interaction between Brefeld's illnesses and his evident accomplishments gave some reason for his barbed arrogance.

Brefeld developed facile methods for studying fungi from single-spore artificial culture, this including efficient sterilization and sequential change from a liquid medium to one gelatin-bodied, and then by the use of agar. In this methodology, he seemingly preceded Koch, pioneer bacteriologist, rather than following him as commonly credited. Nutrient media used were diverse, but manure (dung) was his favorite. He was successful not only in growing innumerable saprophytic fungi but parasitic kinds (e.g., *Ustilago*) in artificial culture.

Using such methods over the years, Brefeld studied thousands of cultures of diverse groups of fungi as a means of elucidating life histories and systematic relationships. Publication ensued in a multivolume *Botanisch Untersuchungen . . . der Mykologie,* the first volume appearing in 1872 and followed by subsequent issuings the rest of his life. The bulk of his contributions to knowledge and theory about the fungi thus considerably exceeded any of the past. Concomitantly, he was a major pioneer in plant pathology as seen from his studies of the mechanisms of infection and his extended work on the cereal smuts.

It is not that everything stood the test of time. Brefeld disagreed violently with his old master de Bary concerning the nature of yeasts, and his insistence on lack of sexuality in the higher fungi soon yielded to de Bary's superior vision. As Brefeld's health deteriorated, he increasingly depended on assistants, and in 1905 parts IX and X of the *Botanisch Untersuchungen* were coauthored. The last volume was published in 1912 after he had been blind several years.

Brefeld made innumerable new observations followed both by prophetic conjectures as well as errors of observation and interpretation. The former outrank the latter to the extent that he can be regarded as one of the founders of both modern mycology and plant pathology. He is commemorated by the genus *Brefeldia* Rostafinki (a myxomycete) and, through orthographic variants, for three other genera.

SELECTED BIBLIOGRAPHY

Anonymous. 1925. Brefeld. Nature 116:369.

Dolman, C.E. 1970. Brefeld, Julius Oscar. In: Dictionary of Scientific Biography. Vol. 2, pp. 436-438, with citations.

Stafleu, F.A. and R.S. Cowan. 1976. Brefeld. In: Taxonomic Literature. Vol. 1, pp. 314-315, with citations.

August Wilhelm Eichler (1839–1887)

Following **de Jussieu's** "natural" classification of phanerogamous plants, two pathways of thought evolved. In simplified account, one is the **de Candolle** to **Bentham-Hooker** to **Bessey** sequence. The other is Endlicher to Eichler to **Engler** (the three e's). Eichler, the middle man of the latter group, provided data and germinal focus from which the Englerian system was developed.

Born in the Germanic state of Hesse, Eichler went to the University of Marburg where, under the influence of a professor by the name of Albert Wigand, he became a reborn botanist. Apparently proceeding no further in formal education, he vaulted to the postdoc life, conventional for those on the way up. Most importantly, he spent several years with Martius (begetter of the ambitious *Flora Brasiliensis*) at Munich whose work he later carried on. Then he was successively professor at Graz, Kiel, and finally at Berlin (1877).

Exceptionally, among the mighty Germans of his time, Eichler is said to have been a modest and unassuming man. He was a careful judge of data and had an exceptional memory for detail. He was hectored by poor health much of his life, eye trouble and leukemia.

His works of most significance for the future were two, a summary *Syllabus* of the Angiosperm families and the more important *Bluthendiagramme* on which he spent fifteen years. The latter remains a source of useful data. It consists of analytical drawings of the flower and inflorescence structure of all Angiosperm families then known. This was neither the first nor the last time this was attempted—it is rather that the quality of Eichler's work exceeded all others. But Eichler was not just a data compiler; the objective was phylogeny, an improved natural classification of flowering plants. He was cautious about speculation and the the meaning of phylogeny; it should not be just a *Schlagwort* (slogan).

Eichler's studies overlapped before and after **Darwin,** whom he vigorously accepted. But Darwin did not really greatly influence the structure of "natural" systems that Eichler and others then postulated. The difference was that there was now a biological rationale on which to base classifications.

Seemingly the initial Englerian *Syllabus* was derived (with some contortions) from Eichler's *Bluthendiagramme.* In certain respects, Engler should perhaps have more carefully followed Eichler. For example, it is said that Eichler used the word *Choripetalae* (including the contentious Amentiferae) to indicate a tendency, not a taxon as later interpreted by Engler.

Eichler's other major accomplishment was that of editorship and author-in-part of Martius' *Flora Braziliensis* after the latter's death, before it was more than one-third completed. (The monstrous Martius multivolume undertaking, indeed, was one imbued with malignant optimism.) Eichler, faithful to his older friend, spent the latter days of his life seeing it through to nearly the end. It remains a much used classic reference. Eichler also contributed the Gymnosperms for Engler's *Das Naturlichen Pflanzenfamilien.*

Evidently Eichler was the initial data and idea person for the Englerian postulates. This led to the Englerian classification system, which dominated all others for half a century. Sometimes posterity overlooks the genesis of concept.

SELECTED BIBLIOGRAPHY

Risse, G.B. 1971. Eichler. In: Dictionary of Scientific Biography. Vol. 4, pp. 306-307, with citations.
Stafleu, F.A. 1965. Eichler's Bluthendiagramme. Taxon 14:199-200.
Stafleu, F.A. 1965. Engler's syllabus. Taxon 14:23-25.
    Includes Eichler as the formative idea person.

Phillippe Edouard Leon Van Tieghem
(1839–1914)

It is sad that a man with so impressive a name should quickly become an orphan, but that is the case. Phillippe Van Tieghem was brought up by relatives. He became probably the best known French botanist of the latter part of the nineteenth century.

For graduate study, Van Tieghem worked with Pasteur on fermentations and obtained the doctorate in physical science. But this was not what he wanted. So he wrote another dissertation and obtained a degree in natural science. He was modern in that he married (1862) before finishing graduate work. There were subsequently five daughters and one son.

Following were a succession of professional posts that are difficult to unravel—either some dates given are incorrect or several appointments were simultaneous. Van Tieghem's innumerable publications (328 by one reference, 357 as given by another) ranged across most of botany, and he usually had several investigations going at one time. Most importantly he was a plant anatomist and comparative morphologist, wherein he made numerous new proposals about tissue homology and origin.

The plant consists of root, stem, and leaf; everything else represents modifications of these fundamental organs. So spoke Van

Tieghem of vascular plants. Of stems and roots, he differentiated between the stele (the name comes from him) and the external cortex. The outer layer of the stele, the pericycle, separates it from the cortex, clearly so in the root, ambiguously in the stem. Van Tieghem kept finding new kinds of steles for which he proposed a complex nomenclature. He differentiated between primary and secondary tissues. He studied the comparative anatomy of the flower in some sixty-seven families. His thesis that the vascular anatomy of flowers reveals relationships brought systematic anatomy into being. Also Van Tieghem was one of the apostles of the appendicular derivation of the inferior ovary, a topic (axial vs. appendicular) that has been the subject of spilled ink since his time. Beyond the flower, he went to ovules and seeds.

Sampling now from work among the "lower" organisms, Van Tieghem was among the first to point to the cellular similarity of the blue-green algae and bacteria. He worked with single spore cultures of fungi but apparently made no major advances in mycology. He investigated butyric fermentation (shades of his past with Pasteur) by *Bacillus amylobacter* and the role of this organism in the natural decomposition of organic material.

Many major botanists have been unable to resist the temptation to frighten unwary undergraduates with a difficult textbook. Van Tieghem wrote several, of which his multieditioned massive *Traité de Botanique* has been best known.

As all to whom we owe our botanical heritage, it is evident that Van Tieghem was a workhorse. He produced a number of new concepts with a minimum of strikeouts. His work suggests that he was a gifted and dedicated individualist.

SELECTED BIBLIOGRAPHY

Green, J.R. 1909. Van Tieghem. In: A History of Botany 1860-1900, pp. 208-217. Clarendon Press, Oxford.

Nougarède, A. 1976. Van Tieghem. In: Dictionary of Scientific Biography. Vol. 13, pp. 405-406, with citations.

Schmid, R. and M. Guédès. 1975. On the Publication of Van Tieghem's "Recherches sur la structure du pistil et sur l'anatomie comparée de la fleur." Taxon 14:659–664.

Johannes Eugenius Bülow Warming
(1841-1924)

Warming, Danish, was "the founder of plant ecology." To ecology he was "father, son, and holy ghost." So it has been said. One can quibble with these absolutes by reciting the names of some of his predecessors, for example, von Humboldt, von Post, Haeckel (who coined the word ecology), and the prose-poetry of Kerner about the plants of the Danube basin. And even **A.F.W. Schimper,** Warming's near contemporary.

I have a multiauthored, 56-page memorial compendium about Warming, published three years after his death (Rosenvinge et al 1927). It would undoubtedly be helpful in evaluating his stature and deeds. Sadly, at least for this reader, it is mostly written in Danish. The following then lacks such understanding as it might bring.

Warming's father was a minister. I have no information about the young Warming's formal education; one presumes he went to college. Neither do I know anything of his personality or marital status. He spent three years as the secretary of P.W. Lund, a paleontologist studying fossil Bradypodidae in Brazil. During this time, he evidently avoided the Bradypods except for execution of whatever formal duties he had. Instead he familiarized himself with the tropical savannah. He subsequently worked with Martius and **Nägeli,** all of this time writing papers and books on a large range of botanical subjects. He became professor of botany at the University of Copenhagen in 1886, a position he held until retirement.

Warming evidently would have been an important botanist even if he had never heard of ecology. One writer says that two books based on Brazilian Savannah studies represent his most outstanding work. Posterity scarcely agrees, but perhaps, like **Darwin's** *Voyage of the Beagle,* they were landmarks on the way. But his research and writing otherwise included *De L'ovule* (1878), which Greene *(History of Botany,* 1909) calls "Warming's great memoir." And a treatise on purple bacteria (1876). And studies of flower biology (1876-1878). And, with the kettle running over, a systematic botany text in 1879 and a general botany in 1880.

For ecology, Warming's *Plantesamfund* (1895) (*Oecology of Plants* in English) quickly became The Book. Concerning it, I am not sure of the origin of the following sentences, taken from some notes of mine several years ago. "Oecological plant geography seeks: (1) to find out which species are commonly associated together in similar habitats; (2) to sketch the physiognomy of the vegetation and the landscape; (3) to answer the questions: why do species congregate to form definite communities, and why do these have a characteristic physiognomy; and (4) to consider the economy of plants and their growth-form."

Warming used water budget and growth forms of plants as the basis of a regional classification of plant associations in North Europe. In doing so, he used the terms xerophyte, mesophyte, heliophyte, psammophyte, and hydrophyte. Duration was defined in terms of monocarpic (flowering once) or polycarpic (flowering repeatedly).

In subsequent reading, I see that **Tansley,** more than fifty years ago, summarized my first paragraph in stating "there were brave men before Agamemnon" (Rosenvinge et al. 1927, p. 54). True, but one must compare the facts about our worthy with those prior. Although one may feel that the contention whether Warming did or did not found plant ecology is beside the point, it seems evident that he was the first to present a unifying concept. And also whether one presently agrees with his concept(s) in toto is beside the point as with Darwin or Freud. The talisman is the word *unify-*

Johannes Warming

ing. In so doing, Warming jigsawed together pieces of both the past and the present in a coherent whole, one which asserted, "Here is a new world!"

The book awoke sleeping dogs by the score. Although floristic plant geography was at best secondary among Warming's percepts, several European phytosociological schools were quickly on the warpath, for example, Braun-Blanquet of the so-called Zurich-Montpellier group. Then came du Reitz' growth form-physiognomy classification and Raunkiaer's overwintering buds: Phanerophytes, Chamaephytes, etc. The message quickly jumped the Atlantic. **Cowles** began studying succession on Michigan sand dunes, and **Clements** was soon developing the dogma of a successional sequence that halted only on reaching Nirvana, the climax.

I've decided not to quibble. Warming founded plant ecology!

SELECTED BIBLIOGRAPHY

Müller, D. 1976. Warming. In: Dictionary of Scientific Biography. Vol. 14:181-182, with citations.

Rosenvinge, L.K. et al. 1927. Eugenius Warming. Bot. Tidesskrift. 39-1-56.
Several writers present their bouquets to Warming; only Tansely is in English.

Warming, J.E.B. 1909. Oecology of Plants. An Introduction to the Study of Plant Communities. Clarendon Press, Oxford. 422 pp.
Groom, P., I.B. Balfour, and M. Vahl, eds. English translation of Plantesamfund, 1895.

Heinrich Gustav Adolf Engler (1844-1930)

Engler, the greatest organizer of the efforts of others in the history of systematics and perhaps of all botany, was already an authority on the local flora when he entered the university. He received his Ph.D. in 1866. A succession of subsequent appointments, typical of a rising comet, ended in 1889 when he ascended to the plum of all, Berlin, professor and director of the Botanic Garden. There, as in all prior positions, he reorganized everything, the garden, the herbarium, the administration. He also traveled widely, especially in Africa, in connection with his own research.

Engler married early and produced two children. I find nothing else about his nonprofessional life (if any) or his personality. His career suggests that he was a successfully domineering Teutonic in the best or the worst sense as one may view it.

I must skip Engler's own writing (numerous monographs, treatises on the plant geography and flora of Africa), with the exception of the *Syllabus der Pflanzenfamilien,* and consider primarily the gigantic, multiauthored productions that he sponsored.

The most influential of these has been *Das Pflanzenfamilien,* a summary of the plant kingdom to the generic level, published 1887-1915. Plant classification therein was summarized at intervals by Engler in the *Syllabus der Pflanzenfamilien* continued by other

Heinrich Engler

authors after his death, the twelfth edition edited by Melchior (1964) being the most recent of these. The impact of these jointly was that of a level of systematic authority unknown before Engler's time except for **Linnaeus,** and certainly not since. Standard manuals that many of the older of us have used (e.g., *Gray's Manual* and the Britton and **Brown** *Illustrated Flora*) follow the sequence of Engler. Until recent years, the filing arrangement in herbaria was that of the *Syllabus*. The Englerian system in fact was holy writ despite the contemporaneous proposals of **Bessey** and others who said that simple does not necessarily mean primitive.

Simple does not mean primitive? No, the opposite. Bessey regarded the buttercup or magnolia-type complete flower, often with numerous pistils, stamens, and petals as the Angiosperm progenitor types (Hallier had similar views about the same time). On the other hand those Engler listed as "primitive" were woody plants with simple flowers—simple meaning lacking some parts, for example, a staminate flower without either corolla or pistil(s). These were the Amentiferae (oaks, willows, birches, etc.), which presently are viewed as including several groups of more or less parallel reductions. But the Englerian bible really did not fade until the 1950s and 60s with the writing of individuals like Cronquist, Thorne, and Takhtajan.

But is this overkill? Did Engler and his colleagues consider their system primarily phylogenetic rather than one of convenience? Engler elaborated his somewhat evolving views of classification several times in *Principien der Systematischen Anordnung* published in the *Syllabus*. A recent analysis of these is that of Barabé and Veith (1990). There the answer is both yes and no, because Engler used two types of series, those believed to be genetic and those morphological. Otherwise, for angiosperms, much of the reasoning and inspiration was derived from the two volumes of **Eichler's** magnificent *Bluthendiagramme* (1875, 1878), which became submerged in the Englerian system. (Eichler was Engler's predecessor at Berlin.)

Anything beyond the *Syllabus* and the *Pflanzenfamilien* might seem prosaic, but that is not the case. I give three examples. Engler started *Das Pflanzenreich* to be monographs of all(!) genera of plants. This impossible project is the only Englerian effort that aborted, but it did extend to some ten feet of shelf space and contains the basic (or the only) reference for revisions of many plant genera. With Drude, Engler (author in part) edited the multivolume *Die Vegetation der Erde*. And he founded a major journal, the *Botanische Jahrbücher,* in 1880 and served as its editor until his death fifty years later.

Engler, let us not forget, was otherwise a major botanical administrator and activist, at least at both Breslau and Berlin. When he came to Berlin, the Botanic Garden and research facilities were cramped within the inner city. He moved them to the outskirts in Dahlem, and the garden became one of the world's botanical showplaces. The herbarium, with its important type-bearing collections (like that of Willdenow, the author of several thousand taxa) dominated all others, except those at Kew, Paris, and Geneva. It was destroyed by Allied bombing in World War II. Happily, the staff had previously secreted the major type collections in the safety of salt mines.

One author has labeled Engler as the The Doyen of both plant geography and taxonomy. I do not quarrel. Also seemingly he was the management Urtype of the nineteenth-century German mass production botanical factories. I find him slightly impossible.

SELECTED BIBLIOGRAPHY

Barabé, D. and J. Vieth. 1990. Les Principes de systématique chez Engler. Taxon 39:394-408, with citations.
Rehder, A. 1937. Adolf Engler. Proc. Amer. Acad. Arts Sci. 71:497-550.
Stafleu, F.A. 1965. Engler's syllabus. Taxon 14:23-25.
Stafleu, F.A. 1978. Engler. In: Dictionary of Scientific Biography. Vol. 15, pp. 147-148, with citations.

Eduard Adolf Strasburger (1844-1912)

I knew of Strasburger when I was an undergraduate student. He was one of those Germans who had a textbook translated into English. I was told in graduate school that Strasburger was the father of plant cytology. Then I forgot about him.

On reading about Strasburger now, I like him. Possibly that egocentric demigod **Sachs** (1832-1897) beat him to the draw in creating the "new botany," but Strasburger's conceptual contributions, lasting through a couple of overtimes beyond Sachs, may have more than equaled the latter.

Though German, Strasburger was born in Warsaw of a mercantile family. His secondary education was in Poland. He was then successively at the Sorbonne in Paris, at Bonn, and then at Jena where he received his Ph.D. (1866). **Pringsheim,** Haeckel, and to a lesser degree, Sachs were his mentors. Robinson (1976) says that Haeckel, a zoologist and **Darwin** exponent, was especially important in providing Strasburger an evolutionary thrust that became the central focus of his otherwise miscellaneous investigations. Evidently he was quickly recognized as a comer because he levitated to a professorship at Jena at the age of twenty-seven (another reference says twenty-five; take your pick). In 1881 he went to Bonn where he stayed the rest of his life. There, he was in charge of the laboratories and the botanical garden. He was also briefly (1891-1892) rector of the university.

Strasburger married in 1870; there were two children. He wrote numerous popular blurbs and a book about how nice it was to vacation on the Riviera. He had numerous students from all over the world, among whom a high proportion came from the United States. Said to have been a man of universal culture, I have the impression that, unlike some of his German predecessors, he was a pleasant and delightful chap.

Professionally, Strasburger was skilled in the details of microtechnique as well as microscopy and was a keen observer. His innumerable publications fall primarily in the areas of cytology (of which he was said to be the founder), morphogenesis, and comparative morphology. Most specifically much of his work related to intrepretation of the reduced gametophytes and sporophyte embryos of the Gymnosperms and Angiosperms. But it is not the bulk of publication or even discernment that makes Strasburger; it is the fact that he possessed a supreme gift of conception. He was able, beyond others, to fit the pieces of the jigsaw puzzle into an evolutionary scenario that made sense. And he wrote a textbook, possibly the most used of all time.

Strasburger's most significant earlier (1875-1880) contributions concerned the details of mitosis. Among several botanists, he seems to have been the primary pathfinder. As early as in his doctorate dissertation he asserted that the nucleus divides during cell division, contradicting the prevalent thought that it only disappeared. Zoologists were making parallel observations, and Strasburger was quickly struck with the similarity of the process in the two kingdoms of life. He could not but believe that these were of common origin, substantiating his prior conversion to fervent Darwinism by Haeckel.

Strasburger meticulously worked out the cellular morphogenesis of vascular plant reproduction. In the Angiosperms he followed the details of the pollen life history from the mother cells to the formation of the generative nuclei, their entrance into the embryo sac, and the fusion of one of them with the egg nucleus. He similarly traced the development of the embryo sac. The only thing he

Eduard Strasburger

apparently missed was the fusion of the second generative nucleus with the polar nuclei. He did parallel work on the Gymnosperms and ferns, correctly confirming and establishing homologies. Except for further details, it was really Strasburger (without named credit) who was presented when I took general botany almost fifty years later.

Strasburger quickly grasped the idea that chromosomes must represent the hereditary units derived from past generations. Maintained in every cell, they in turn maintained the "like begets like" to subsequent generations. Thus, in fertilization, he said that only the chromosome material of the egg and the sperm were of significance in inheritance.

In the 1890s, Strasburger was one of those who worked through the details of reduction division, and the terms haploid and diploid come from him. He quickly became aware of the difference in chromosome numbers between the alternating generations, not only in the Archigoniatae, but in miscellaneous algae and fungi. In the then heated flames of the antithetic vs. homologous theses of the alternation of generations, this might seem to be fuel for the proponents of the homologous idea, but Strasburger seemingly never made an issue of this. To the contrary, perhaps the varied and sometimes hit-and-miss nature of this alternation among the Thallophytes suggested to him that it might be entirely independent of that exhibited by the "higher" groups.

Beyond these highlights of what we today would call cell biology, Strasburger's investigations extended to various algae and the Myxomycetes. He also published on paleobotany and plant physiology. Enroute, the terms chloroplast and cytoplasm are attributed to him. His *Lehrbuch* (textbook), in numerous editions, was one of the most influential and widely used ever written. Indeed, it has continued to the present under his name by a succession of authors and in several languages.

Morton (1981) finishes his chapter, "The Foundations of Modern Botany," with an appropriate epilogue: "It is fitting to close our account . . . with the death of Strasburger. This modest, cultivated man . . . left his impress on modern botany . . . through his technical mastery, his philosophic reflectiveness of thought, his unswerving scientific integrity, and a charm and humanity to which countless students throughout the world responded with grateful affection."

And indeed Strasburger was the supreme graduate student proctor, and his numerous apostles continued his tradition beyond his immediate life, for example, among those in the United States, Campbell, Chamberlain and Duggar. As a major river that flows into the sea, its contribution is quickly diffused and its identity soon fades. But among rivers, Strasburger was at least a Mississippi or perhaps more appropriately, a Rhine or Danube.

SELECTED BIBLIOGRAPHY

Jackson, B.D. 1912. Eduard Strasburger. Proc. Linn. Soc. London. 1911-1912:64-67.
Morton, A.G. 1981. Strasburger. In: History of Botanical Science, pp. 435-440. Academic Press, London.
Robinson, G. 1976. Strasburger. In: Dictionary of Scientific Biography. Vol. 13, pp. 87-90.

Charles Edwin Bessey (1845-1915)

Bessey was probably the most important U.S. botanist west of the Mississippi from maybe 1880 until 1910. In fact, following the demise of **Asa Gray** (1888), this might be true for the whole country.

Our subject was born in a log cabin on his parents' farm near Wooster, Ohio. His formal secondary education was sporadic. At the age of twenty, he went to Michigan and worked as a surveyor, a timber cruiser, and a schoolteacher. In 1866 he enrolled at Michigan State College in civil engineering but subsequently changed to plant science. President Abbott reportedly said, "Well Bessey I am glad of it, but you will never be rich." After graduation, he came to the recently established Iowa State College as an instructor. Michigan State subsequently (1872) awarded him a master's degree, whereupon Bessey was promoted to professor of botany, horticulture, and zoology at Iowa State. His duties, however, rapidly became diverse even beyond this elongated title, and he once remarked that he occupied not a chair, but a settee.

Asa Gray, the reigning monarch, invited Bessey to come to Harvard to study with him, and Bessey spent three winter breaks (a month each) with the master. But he did not pursue further graduate work, his several subsequent doctoral degrees being honorary.

Bessey left Iowa State for the University of Nebraska in 1884, presumably as a consequence of a dispute between the college administration (Bessey was vice president in 1883) and the legislature. He remained at Nebraska as a member of the faculty and administration the rest of his life.

Bessey was a short, stocky man who was an incessant bubbling conversationalist (in a "heavy, round, cheery voice") especially on plants, but including any other subject whatsoever. He married Lucy Athearn in 1873. They had three sons, one of whom, Ernst Bessey, became one of the better known American mycologists of the first half of the twentieth century.

Professor Bessey's activities and accomplishments remain fatiguing to behold. American education in botany, when Bessey came on the scene, was either that of a handmaiden to the medical profession or a gentle and innocent book subject especially for young ladies. Bessey changed that. He said a laboratory was as essential in botany as in chemistry. By 1873 there was a formally scheduled laboratory course at Iowa State that included the use of the microscope. It was the first such undergraduate course in the country. And furthermore, Bessey was an evangelistic teacher who drew students by the droves. The master's degree (1878) of his first graduate student, J.C. Arthur (later a renowned rust mycologist), was also the first awarded at Iowa State.

At Nebraska, the same pattern repeated. There were initially no microscopes, and Bessey stipulated their purchase as a condition of his moving to that institution.

Until Bessey's time, formal education had little to do with agriculture. Agriculture was an independent art, passed from father to son, often little changed for many generations. Its first substantive entrance was in the "schools for the people"—the land-grant institutions in the United States. Bessey pioneered in making "for the people" a reality. He was a one-man extension service; for example, he tested the usefulness of more than 100 varieties of vegetable crops under Iowa conditions and reported about insect pests and diseases in lectures and popular writing. He initiated agricultural research both at Iowa State and Nebraska. He established a Farm Institute in Iowa as an organization essentially undertaking what we today call extension education. With the cooperation of Governor Furans of Nebraska he organized similar institutes in Nebraska. And at the national level, he helped draft and promote legislation

Charles Bessey

leading to the establishment of federally funded experiment stations in each state; i.e., the Hatch Act that was passed in 1887. He was the first director of the Nebraska Experiment Station. His biographer Walsh (1972) has said, "Charles Bessey is the epitome of the land-grant professor." True, ideally; but Bessey was more nearly a unique land-grant professor.

Bessey's promotional-evangelical nature found further expression in his role as an editor and as an officer of scientific societies. He made the most of his editorial prerogative, especially for the journals *American Naturalist* and *Science*. He wrote hundreds of book reviews as an exponent of the "new botany" of the burgeoning Germans, a botany that went beyond observation to experimentation and conceptualism. He was a prodigal organization man. One of the founders both of the Iowa Academy of Science and that in Nebraska, he quickly became president of both and likewise of nearly any other organization he joined. Here I note only the Botanical Society of America (1897) and the American Association for the Advancement of Science (1911).

And Bessey was an early conservationist; he was involved in the establishment of the Nebraska State Park Commission, an early movement to save the *Sequoiadendron* (giant redwood) groves in California, and member of a committee to establish the Yellowstone Reserve. In an entirely different area, he was active in the improvement of secondary education and was vice president of the American Education Association one year.

Amid all of this, one wonders if Bessey also had time to be a botanist. Amazingly, yes! Among the clutter of short journal notes and book reviews that he poured out all of his life, one can count two major achievements that left a permanent mark on posterity.

Bessey followed Asa Gray as the foremost textbook writer in the United States. These extended from the first edition of his *Botany for High Schools and Colleges* (1880, the first of four editions) through a sequence of titles to his *Essentials of College Botany*, coauthored with his son Ernst A. Bessey (1914). Strongly influenced by **Sachs'** multieditioned *Lehrbuch der Botanik* (1870), these texts, according to Ewan (*A Short History of Botany in the United States*, p. 46), "reoriented botanical instruction in this country."

And finally the classification of flowering plants. During the middle and latter years of Bessey's career, the multivolume, multiauthored German compilation of *Das Pflanzenfamilien* (**Engler** and Prantl) was the reference bible for the flowering plants. The arrangement of families in various editions of the *Syllabus* of this German monster was that used in nearly all manuals and herbaria. Presumatively it tabulated once and for all the probable relationships among plant families. Most specifically, it held that plant groups with "simple" flowers (the Amentiferae) without calyx or corolla (or lacking both), which were often unisexual and wind pollinated, were the primitive (plesiomorphic) kinds. Those with complete flowers were derived (apomorphic). In a couple of preliminary papers and in the finale, *The Phylogenetic Taxonomy of Flowering Plants* (1915), Bessey upended all of this. On the basis of a series of explicit dicta, he asserted that the plants with bisexual flowers with numerous petals, sepals, etc., were the basic types from which the reduced kinds, for example, willows, oaks, and birches, representing several unrelated but superficially similar lines, were derived. Botanists were interested in Bessey's revolutionary proposals (and a similar one from the European botanist Hallier) but serious restudy of higher plant phylogeny did not come for some years. Then it abruptly arrived of age in the 1950s and 60s primarily with the productions of Cronquist, Thorne, and Takhtajan. Although the systems of these authors differed in details, that of Bessey was the direct and common ancestor of all. Cronquist's book carries a frontispiece photograph of "Charles E. Bessey, author of a phylogenetic system . . . which profoundly influenced subsequent taxonomic thought."

SELECTED BIBLIOGRAPHY

Mertins, C.T. and D. Isely. 1981. Charles E. Bessey: Botanist, educator and protagonist. Iowa State J. Res. 56:131-148, with citations.

Overfield, R.A. 1975. Charles E. Bessey: The impact of the "new" botany on American agriculture. Technology and Culture 16:162-181, with citations.

Overfield, R.A. 1993. Science with Practice: Charles E. Bessey and the Maturing of American Botany. Iowa State University Press, Ames. 262 pp., with citations.
Overfield's recent book includes insights about Bessey beyond those in this short essay.

Pammel, L.H. 1928. Prominent Men I Have Met: Dr. Charles Edwin Bessey. Ames, Ia. 20 pp.

Walsh, T.R. 1972. The American green of Charles Bessey. Nebraska History 53:35-57, with citations.

William Fredrich Philipp Pfeffer
(1845–1920)

Sachs and subsequently Pfeffer (both German) were the duo who established modern plant physiology. Both wandered all over the field, wrote texts, and had innumerable students from everywhere. Nearly all of the first-generation American plant physiologists were trained by one or the other.

Pfeffer was fourth generation in an apothecary family, and most of his formal training was directed towards continuing the tradition. But as a young man he was also interested in mountain climbing (one of the first to scale the Matterhorn) and in plant collecting and classification. His ultimate career seemingly derived from his interest in plants and was supported by extensive training in the physical sciences. He obtained a Ph.D. in botany and chemistry in 1866 at age twenty. He was now enamored of an academic career, plants for plants, not as diverse cure-alls for humans.

Pfeffer was successively professor at Bonn, Basel, Tubingen, and finally at Leipzig in 1887. There he was director of the Botanical Institute, which it is said had facilities for twenty advanced students at a time, i.e., extensive laboratory space and equipment and a research garden. The top floor of the building constituted living quarters for the director. Pfeffer stayed there the rest of his life.

Marriage came in 1884. One can say that Pfeffer took it seriously, because after that time he stopped participating in the more dangerous mountain climbing. One son, who came to the union, was killed in 1918 in the war.

The reasons for Pfeffer's preeminence included not only his gifts of mind and spirit but his training in both chemistry and physics. In these fields, his competence apparently exceeded that of any major botanist of his time. Furthermore he was not only an outstanding lecturer but a laboratory research teacher as well. The combination of the man and the facilities brought students from all over the world.

Pfeffer's own research covered the waterfront. His studies of osmosis are classic. He developed methods for measuring osmotic pressure and explained its rationale in the language of physical chemistry. He devised artificial so-called "pepper pot" osmotic cells from treated, ordinary glazed planting pots. It is said the van't Hoff's work on osmotic and gas pressure owed much to Pfeffer's contemporary investigations.

As to the scope of his work otherwise, I can only give a partial census enumeration. Pfeffer studied the mechanisms of "sleep movements" in plant leaves; also, the opening and closing of flowers. With the sensitive plant (*Mimosa pudica*) as model, he investigated response to physical stimulation. He compared the efficiency of various light spectra on photosynthesis. He wrote on respiration and protein metabolism. He opened the world of chemotaxis, for example, showing that malic acid is responsible for movement of sperm cells in ferns and *Selaginella* to the archgonia. In these and other studies, his superb mental gifts, his skill in instrumentation, and knowledge of the physical sciences placed the level of his work above most others.

Pfeffer, of course, had a handbook—what German could be without it—and some such books often required two hands to carry. His *Pflanzenphysiologie*, two editions, were translated into English. Both a reference work and a guide for the future, it was, so it has been alleged, "the greatest production of its kind." Aphoristically the author said: "Changes in energy control functioning; when energetics is understood, one will comprehend life."

When World War I came, Pfeffer was on top of the world in

his profession, hailed by the past and the present alike. A vast celebration, planned for him on occasion of his seventieth birthday, however, was hampered by the war. And sadly, Pfeffer, understandably depressed by the destruction of his country, was deeply unhappy his last years.

A 1920 obituary in the English *Nature* said, "With his death, the three outstanding figures of the older German botany—Sachs, **Strasburger,** and Pfeffer—have all passed away. Indeed the end of an era."

SELCECTED BIBLIOGRAPHY

Andrews, F.M. 1929. Wilhelm Pfeffer. Plant Physiology 4:285-288.
G.J.P. 1920. Wilhelm Pfeffer. Science 51:291-292.
Robinson, G. 1974. Pfeffer. In: Dictionary of Scientific Biography. Vol. 10, pp. 574-578, with citations.

Hugo de Vries (1848-1935)

To the extent that anyone remembers anything from a general biology course, the name de Vries may possibly be retrieved from the mental rubble. Perhaps one may also recall that de Vries was the Mutation Man. That is a reasonable summary.

De Vries was Dutch. His formal education moved sequentially from the Netherlands (Leiden) to Heidelburg, Germany, and then to **Sachs'** laboratory at Wurzburg where he received his doctorate in plant physiology. He subsequently returned to his own country at the University of Amsterdam in 1871 and was professor and director of the botanic gardens until formal retirement in 1918.

I must overlook de Vries' initial work, primarily in physiology, especially of the physical chemistry of osmotic pressure, to better fill in the mutation story. For, mid-career, he turned primarily to the prevalent biological excitement of the time, **Darwin's** evolution by natural selection among a continuum of small variations. But Darwin was not able to explain the origin of differences between species or individuals of plants and animals, nor their mode of inheritance. Recognizing this, de Vries and some others proposed a theory called pangenesis: namely that innumerable particles, molecular pangenes from all parts of the body, moved to the sex cells, there determining the features of the next generation. The idea remained at best murky until the latter nineteenth century, when

Hugo de Vries

knowledge of the chromosomes and their behavior during reduction division provided a locale for the pangenes and a possible supporting mechanism. (The word gene is derived from de Vries' pangenes.)

But this proved to be just background for de Vries' major proclamations (1901-1903). Speciation is not a result of the additive effect of small variations, he said. Rather, it derives from sudden jumps that he called mutations. His observational and experimental data came primarily from the evening primrose (*Oenothera lamarckiana*). He found saltations, parents to progeny, among these plants and saw that the new types, species in his mind, were now consistent in their characters and bred true. The mutation theory became one of the major motifs that genetic theorists had to support or unravel during the next thirty years.

De Vries was one of three individuals who more or less simultaneously and independently discovered **Mendel's** work. The good monk's data, however, seemingly placed a damper on the mutation thesis. So de Vries subsequently tried to have his cake and eat it too; he acknowledged Mendel, but simultaneously maintained that the mutations were more important as the basis of speciation and that these new taxa were the units that nature would try out in Darwin's natural selection.

But as many of you know, de Vries' mutations were not that at all in the sense that he initially interpreted them, i.e., gene mutations in our language. As has subsequently been demonstrated (especially by Ralph Cleland of Indiana University, 1972), they were the consequence of miscellaneous chromosome hanky-panky, polyploidy, trisomics, and reciprocal translocations resulting in ring formation rather than chromosome pairing at meiosis.

Thus de Vries' mutants lack the indelible evolutionary significance that he attributed to them. Had he waved a flag where nothing really existed? So it seems. But his work provided an extraordinary stimulus to evolutionary geneticists who were seeking alternatives to Darwin's gradualism through the wicket of Mendelian inheritance. I guess that is pretty well settled now. But a seminar title I recently saw was entitled, "Why genes in chloroplast and mitochondria don't obey the laws of Mendelian genetics, and some evolutionary consequences."

SELECTED BIBLIOGRAPHY

Cleland, R.E. 1972. Oneothera Cytogenetics and Evolution. Academic Press, London. 370 pp.

Farrall, L.A. 1973. de Vries. In: McGraw-Hill Encyclopedia of World Biography. Vol. 11, pp. 191-194.

van der Par, P.W. 1976. de Vries. In: Dictionary of Scientific Biography. Vol. 14, pp. 95-105, with citations.

Vries, Hugo de. 1909-1910. The Mutation Theory: Experiments and Observations on the Origin of Species in the Vegetable Kingdom. 1 vol. Open Court Publishing, Chicago, Ill.

English translation by J.B. Farmer and A.D. Darbishire.

Luther Burbank (1849–1926)

Burbank was a seat-of-the-pants, pioneer plant breeder who became publicly notorious during his lifetime and afterwards. That his fame (or infamy) was not ephemeral is evident in the spate of books about him that have appeared in the last century. He has been called a genius, a wizard, a mystic, an infidel, a magician, and a victim of adulation, and some of this seems true. I list him here as a botanist because the supporting kingpin of all this was his work with plants.

Burbank was born on a farm near Lancaster, Massachusetts, his father's thirteenth child by his third wife. After attending a college preparatory academy, he terminated formal education, taking a menial job with a plow company. Three years later, he succeeded in obtaining a small farm where he could do what he wanted—to make a living by growing vegetables and developing better ones. The incentive possibly derived from his youthful reading of **Darwin's** *Variations of Animals and Plants under Domestication.* (**Mendel** was not yet on the scene.) And he came up with something new, the Burbank potato, immediately accepted in Ireland still recovering from the scorge of the potato blight. The subsequent half-life of the Burbank has been much extended, inasmuch, so I am told (J. Niederhauser, personal communication), that it is the Burbank potato that is used by the McDonald's fast-food empire.

In 1875, Burbank moved to California seeking a climate more conducive to his work. He successfully established himself as a farmer and nurseryman, and in a few years was devoting nearly full time to his primary passion, that of changing plants at hand to something else better or more attractive or interesting to humans. This was his life thereon except for two marriages. It was not that Burbank rushed into holy matrimony. He waited until age forty-one to try the first time. Worse luck! Lasting but six years, the union was a dismal, even spectacular failure, ending with Burbank sleeping in the barn (Kraft and Kraft 1967, pp. 62–67). Burbank waited it out for twenty more years before daring again. He then married his secretary; he was sixty-seven and she was said to be in her twenties. Here perhaps an unlikely coupling, but successful. Several references say that Burbank liked children, but there were none of his own.

Returning now to the primary plot, Burbank developed new kinds of innumerable plants (fruit, vegetables, and ornamentals) by making crosses and following them through subsequent F_1, F_2, etc., generations, looking for characters he was after and eliminating those deemed undesirable. Often he included genetic material from several species. His methodology seemed to be intuition rather than science. To move more quickly in evaluating progeny of woody plants, he grafted his seedlings on to established rootstocks. He developed new varieties of *Prunus* (plums and cherries), new *Rubus* (raspberries, blackberries, and their kin), a giant rhubarb, the Shasta daisy, new roses, new sweet corns, a *Pisum* (pea) especially for canning, a giant artichoke, even a spineless cactus (*Opuntia*?) said to be a good forage plant for arid areas. All of this plus the fact that Burbank had views on everything and made good newspaper copy. He drew national and international attention, which brought the epithets wizard, genius, and the like. Burbank included among his friends Henry Ford and Thomas Edison—birds of a feather I suppose.

Burbank was a "free thinker" who bluntly rejected the divinity and miraculous conception of Christ. "There was no proof of it," he said. From this comes the term "infidel" used in the titles of two books about him. But he was not consistent. He was a mystic in terms of paranormal communication. He believed he could sum-

Luther Burbank

mon his sister who lived near by and who was somewhat of a caretaker during most of his bachelor years by thinking about her. Inasmuch as he could do things with plants using about the same procedures as others who failed, he believed, so he said, that he was a mystic with plants; he talked to them, assuring that all would be well if they followed his directions. He passionately felt that the human race needed improvement (who can quarrel with that?) and his eugenic proposals were a transposition of his plant breeding credos—let only the fit reproduce the kind.

It is not that Burbank's life was one entirely of adulation and success. It is hard to trace his experimental methodology because of his hit-and-miss record keeping. Some scientists attempting to duplicate his productions failed and labeled him a fraud; some journalists eagerly took this up. His work was supported for five years by the Carnegie Foundation (the National Science Foundation substitute of its time), but the contract was not renewed because the foundation was uncertain of what it was getting for its money. The double role of marketing what he had and developing what he did not have led to periodic episodes of financial crisis. There was no real way of copyrighting his products. No doubt he lacked the creative approach to business that he had in the improvement of his merchandise.

What to make of this man who, with **George Washington Carver,** both marginally scientists in the strict sense, probably drew more U.S. public attention than any others who worked with plants. Burbank was not the first to cross plants and find something different in the progeny. Koelreuter, Mendel, and others had done that many years before. But Burbank jumped in where there was yet a void in practice and made the most of it; perhaps he indeed was a genius in his ad lib interpretation of his breeding results. He was a good man (in a humanist, ethical sense), who was capable of putting his views on anything into pungent aphoristic phrases that caught attention. Perhaps he indeed was a developer rather than a scientist per se, but he opened the eyes of the more conservative often skeptical scientists who in Burbank's last twenty years had the advantage of Mendelian genetics.

I am glad to claim Burbank as a worthy botanist who used his gifts to graft action with his vision.

SELECTED BIBLIOGRAPHY

Crosby, J.L. 1946. Burbank. The New Phytologist 45:289-290. Review of W.L. Howard, 1945-46. Luther Burbank. A victim of hero worship. Chronica Botanica 9 (5-6):299-522.
Kraft, K. and P. Kraft. 1967. Luther Burbank: The Wizard and the Man. Meredith Press, N.Y. 270 pp.
  Perhaps the most carefully researched and well written of the Burbank books.
Lewis, J. 1930. Burbank: The Infidel. Free Thought Association, N.Y. 29 pp.
  A somewhat slimy epistle focusing on one topic.
National Cyclopaedia of American Biography. 1947. Burbank, Luther. Vol. 33, pp. 149-150. James White and Company, N.Y.

John Merle Coulter (1851–1928)

J.M. Coulter and **C.E. Bessey** were probably the best known midwestern U.S. botanists during the latter part of the nineteenth and early twentieth century. Maybe I could leave out the "midwestern." Coulter started as a systematist, thereafter becoming a jack of most botanical trades, editor, administrator, and eloquent public spokesman for all things botanical.

John Coulter was born in China, his father being a Presbyterian missionary who died soon after the son's birth. Mrs. Coulter brought her family back to their familiar Indiana origins and successfully monitored them to adulthood.

Coulter graduated from Hanover College, Indiana, in 1870, was then a teacher of Latin in a girls' seminary, Logansport, Indiana, then went back to Hanover College as a Latin professor. He succeeded to the chair of natural science in 1874.

It was still the time of exploration of the yet expanding United States. Consequently, Coulter was able to obtain a position as assistant geologist on the Lt. Hayden Exploring Expedition of the Yellowstone and Colorado areas. He rapidly became "The Botanist." In the process of working up the collections, he corresponded with others of his ilk including the venerable **Asa Gray** with whom he subsequently spent several summers. In 1879, he moved to Wabash College, Crawfordsville, Indiana. Apparently he quickly showed management talents, because after receiving the doctorate, he became president of Indiana University. In 1893, he moved to the presidency of Lake Forest University, Chicago. After three years, however, he had his fill of fund-raising and bureaucracy. He resigned and became once more a botanical soldier as the head professor of botany in the nascent University of Chicago. He stayed there until retirement (1925). But he didn't retire. He then accepted a position as chief scientific adviser for the newly founded Boyce Thompson Institute and moved to Yonkers, New York. Vigorous to the end, he died three years later of a heart attack.

Professionally, Coulter was magnetic in personal contacts and an eloquent missionary in public appearances. He was exorbitantly in demand as an invited speaker for scientific meetings, commencements, seminars, and the like. At Chicago, he and the faculty, which he assembled, attracted graduate students by the dozens. The Coulter days were the Chicago glory days in botany.

But scientific accomplishments? From the stimulus of the Hayden expedition, he started out as a floristic systematist, authoring, usually in collaboration with others, pioneer floras of Colorado, the Rocky Mountains, and West Texas. He was junior author with Sereno Watson of the sixth edition of Gray's *Manual* (1890). Then moving toward monograph specialization, he and his student, H.N. Rose, published extensively on New World Umbelliferae. After coming to Chicago, the specialty of his numerous protégés became comparative and developmental morphology and life history studies. This was the exciting time of elucidation of the reduced gametophyte of the Gymnosperms and Angiosperms.

Coulter followed Gray and then **Bessey** as the most important textbook writer in the United States. A succession of elementary texts led to the two-volume *Textbook of Botany,* Coulter, Barnes, and **Cowles** (1910). Coulter wrote the part on morphology. It was almost the standard advanced text for some years.

The Botanical Gazette is presumably well known to all current botanical readers. Its origin traces to Coulter in 1875, and he remained as editor for half a century.

An acclaimed star invariably draws detractors. Shinners in a certainly litigious paper (1963) pointed out correctly that nearly all of Coulter's books were of joint authorship and branded Coulter as an operator, a self-serving mediocre botanist who put his name on any worthy paper or book by any one of his students or colleagues. Well yes, a man who has spent part of his life as a university president is likely to be an operator. No doubt Coulter's name appears in some publications in which he was primarily an enzyme. If so, what an enzyme he was. He certainly was not a mediocre botanist, and I am willing to forgive him.

The passage of time dims the memory of any of us except (sometimes) for publication. Beyond publication, Coulter was an inspiring teacher and an unexcelled tub-thumper for all things botanical. We could use more of his kind now!

SELECTED BIBLIOGRAPHY

Cowles, H.C. 1929. John Merle Coulter. Botanical Gazette 87:211-217.
Fuller, G.D. 1929. John Merle Coulter. Science 69:177-180.
Rodgers, A.D. III. 1944. John Merle Coulter. Princeton University Press, Princeton, N.J. 321 pp.
Shinners, L.H. 1962. Evolution of the Gray's and Small's manual rangers. Sida 1:1-31.

Marcus Eugene Jones (1852–1934)

Marcus Jones, a phenomenon of nature, was an unsurpassed botanical collector in the western United States, an excellent, albeit choleric, taxonomic botanist and a dynamic, inflexible individualist. Most of his work was done outside of the "structure" for which he primarily had contempt and derision.

Jones was born in Ohio. His childhood years were spent mostly on a farm near Grinnell, Iowa. He graduated from Grinnell College with a classical education (1875) and subsequently taught Latin there for three years, studied science, and obtained an M.S. He collected plants in Colorado (1876-1877) and in 1879 became professor of natural science at Colorado Springs. He married in 1880 and almost simultaneously, with his new bride, Anna Elizabeth Richardson, moved to Salt Lake City, Utah. A week after arriving there, he took off for an extended trip in southern Utah. Adams (1938) says that Mrs. Jones accompanied him on this first post-marriage trip—nobody else mentions her. Be that as it may, staying home with eventually three children seemed to be her primary role for life. She was a single-parent mother also with fiscal responsibility for the family; the support Jones obtained was used for his travels and attendent writing (Broaddus 1935). Anna Jones died in 1916.

Jones seemingly never held a continuing salaried position. A

Marcus Jones

jack of many trades, he was a mining consultant and vice versa, the star witness for smoke pollution suits against mining companies (e.g., Anaconda copper in Montana), whose output from ore smelting devasted everything for miles about. Jones also had short-time support from the United States Department of Agriculture and extended help from a General Carpenter, Jones having come to the latter's attention through finding and returning a purse or billfold that the general had lost (an honest man, b'god). He also derived some income from the sale of duplicate specimens.

Jones' personality was something else; he was multifaceted—brilliant, egocentric, obsessed with his turf (the U.S. Great Basin), and almost paranoid about anyone who dared to invade it. He was a loner and devastatingly critical of others, for example, Tidestrom, Rydberg, Heller, Aven Nelson, and **Britton** and his American code of nomenclature. (Jones had his own Code of Nomenclature.) And even E.L. Greene, who was also sufficiently diversive that he had to publish beyond the eastern botanical establishment, drew his wrath. Greene's death brought an obituary statement by Jones that started, "Greene, that pest of systematic botany, has gone and relieved us of his botanical drivel." And of the professional inhabitants of botany in general he stated that he would be glad to subscribe towards a fund for the education of professors, "so that they could go to a botanical kindergarten."

On the other hand, Jones was a fundamentalist Sunday school teacher who said that he never collected a plant on Sunday. He was parsimonious; he touched neither tobacco nor alcohol and ate plainly and frugally. He did much of his traveling in an ancient Model-T Ford that went through innumerable rebuilding jobs and in which he was killed in California in 1934 returning from a field trip.

In 1923, Jones sold his herbarium to Pomona College in southern California, probably because its maintenance was now beyond his means, and because he received the then extraordinary sum of $25,000 for it. And more or less simultaneously, he moved to Claremont, California, likely to be near his unsurpassed research treasure trove.

But, why do we devote space to this eccentric? It is not that he was by far the most prolific collector of his region—not even because his plants included many entities not hitherto known to science. (During the forty-five-year period, 1843-1879, about one-third of *all* Utah-type specimens were of his gathering.) It is rather because he was also a prolific publishing author. From his collections, he described some 165 new taxa (Welsh 1982), and most of his species, unlike those of the microvisioned Greene, have stood the test of time. He was not only a good botanist but one who published extravagantly.

How he accomplished this, far from any good botanical libraries and herbaria until he moved to California, and also cut off, apparently by edict, from the conventional eastern outlets for publication, is slightly miraculous. For the first, I have no satisfactory answer, except that over the years he did indeed visit other major herbaria to see the collections of others; otherwise, he possibly felt that he, himself, was a sufficient library. The second is simple. He started his own private journal, *Contributions to Western Botany*, that went through some eighteen volumes, terminated only by his death. He bought typesetting equipment and personally did the typesetting and otherwise financed his publication. There his new species and botanical observations were recorded.

Except—this was exclusive of his main event, and perhaps the others were small potatoes compared to it—a monograph of North American *Astragalus*, probably the largest genus of flowering plants in North America, 288 pages plus 78 plates, all typeset by himself. Although Jones' individualist and frenetic style, which would have been tolerated in no edited journal, was an impediment to its general usefulness, it is the major podium behind the superb treatment of Barneby published some forty years later. Barneby (1964, pp. 6-8) has presented the best balanced judgment:

> Jones' Revision initiated a new era in the systematics of *Astragalus* . . . it is nevertheless an exasperating work . . . to the novice it is impenetrable, because the path to a given species lies through a labyrinth of polychotomous keys, but to the advanced student who has learned where to look, it is richly rewarding, . . . this is often obscured through Jones' volatile impatience . . . expansive, dogmatic and diffuse, . . . symptomatic of a forceful and choleric personality, . . . traveling as he did year after year, Jones came to know his species

in a way impossible for a museum botanist . . . his opinions have almost all been vindicated by subsequent scrutiny.

The first sentence in this essay states that Jones was a phenomenon. That, I believe, is validated. He was not just idealistic but a cantankerous, collector-pioneer. He was an important North American systematic botanist. More than that, a reader does not need to be a botanist to enjoy his engaging travelogues that are included in most of the issues of the *Contributions,* nor his observations about other botanists. And he did not hate them all, as clearly portrayed in the chapter, "Botanists Whom I Have Known" in the last volume of the *Contributions.*

Jones is commemorated by the legume genus *Jonesiella,* and almost innumerable species and varieties of flowering plants bear the epithet *jonesii.* Saluto.

SELECTED BIBLIOGRAPHY

Adams, T.C. 1938. Marcus E. Jones. Utah Acad. Sci. Arts and Letters 15:11-13.
Barneby, R.C. 1964. Jones. Atlas of North American Astragalus. Mem. N.Y. Bot. Gard. 13:6-8.
Broaddus, M.J. 1935. Marcus E. Jones, A.M. Contr. West. Bot. 18:152-157. (Mabel Jones Broaddus was a daughter of M.E. Jones).
Stafleu, F.A. and R.S. Cowan. 1979. Jones, Marcus Eugene. Taxonomic Literature. Vol. 2, pp. 456-457, with citations.
Welsh, S.L. 1982. Jones. Utah plant types—historical perspective 1840-1981—annotated list and bibliography. Great Basin Naturalist 42:129-195, with citations.

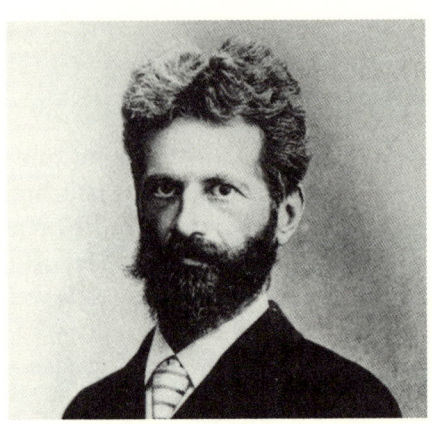

Gottlieb Haberlandt (1854-1945)

Haberlandt, Austrian, is said to have had "great interest and talent in music, painting, and German literature." So he became a botanist. He was an anatomical physiologist or a physiological anatomist or otherwise as you may prefer. Historically, perhaps one reason Haberlandt is known is because he did so many things. Among them, I am not sure which were (or are) of most nominal importance—accounts about of his accomplishments differ widely. Perhaps in total assessment it is that the sum total of ten ones add up not to ten but twenty.

Obtaining his Ph.D. in 1876, Haberlandt then undertook a sequence of postdoc equivalents, the most significant of which was with Schwendener at Tubingen. After some years he accepted a professorship at Granz. In the meantime Schwendener had gone to Berlin and upon retiring, suggested that Haberlandt succeed him. Haberlandt did so (1910) and there remained until retirement in 1923. (The professor evidently had a better chance to pick his successor than now.)

Haberlandt was twice married, five children by the first wife, two by the second.

It is said that Haberlandt had a research program planned while yet in school. This modestly was to investigate all twelve (by his count) plant tissue systems and to relate function to morpho-

Gottlieb Haberlandt

genesis and ultimate structure. He indeed moved from one to another, the photosynthetic system, the apical growth of phanerogams, assimilation, conduction, etc. These and those to follow were the Vorarbeiten (preparatory material) for a Germanic-style 10-pound handbook. But he saw that if he waited until finishing the whole lot that the handbook might never come. So he interrupted to prepare *Physiologische Pflanzenanatomie* (1884). It went through some six editions and several translations. His ideas and the results of some of his research, however, did not meet with unanimous applause. **De Bary** termed the handbook as the "newest botanical novel," and it is also related that some professors did not want their students to read it for fear of leading them down pathways of intellectual sin. But at a minimum, the book established Haberlandt as one of the heavyweights in physiology, one with prophesy for the future.

And Haberlandt continued, for example, working on cell division, wound hormones, and cell and tissue culture. Although he was one of the first to attempt tissue culture, he was successful in only a preliminary way, and subsequent judgements about the significance of his work are not concordant. One author who details Haberlandt's life accomplishments does not even mention tissue culture. Presumably he thought it to be too trivial. On the other hand, a paper in the *Botanical Review* (*Plant Cell and Tissue Culture: The Role of Haberlandt*) is devoted specifically to this topic (Krikorian and Berquam 1969).

A former student of Haberlandt's wrote a tribute to Haberlandt (Noé 1934). Why, Noé asked, did the Germans produce so many intellectual leaders of their time? "Why is it that we now have so many more scientifically trained men, but a smaller number of those of brilliance?" It was, he said, because a humanistic education created the intellectual leaders: "The greatest change in the history of civilization is the disappearance of the classical education." Without disagreeing with the merits of a humanistic education, this latter statement seems a bit overblown. As a matter of fact, I disagree with the assumption about the proportion of brilliance among scientists. Perhaps there are now several or numerous Babe Ruths and none of them stand out as he did in his time. But I do not quarrel with the Noé thesis that Haberlandt probably "belongs to the intellectual aristocracy of the classic era."

SELECTED BIBLIOGRAPHY

Krikorian, A.D. and D.L. Berquam. 1969. Plant cell and tissue cultures: The role of Haberlandt. Bot. Review 35:59-67.
Noé, A.C. 1934. Gottlieb Haberlandt. Plant Physiology 9:851-855.
Sattler, R. 1972. Haberlandt. Dictionary of Scientific Biography. Vol. 5, pp. 623-624.

Dukinfield Henry Scott (1854-1934)

There have been and are those haunted by the ghosts of plants long gone. But it was not until the latter half of the nineteenth century that their kind began to multiply and paleobotany to flourish. Our subject here was a leading synthesizer of paleobotany and an important botanist otherwise.

Scott, English, inherited wealth following the death of his father, a celebrated (and presumably expensive) architect, and was spared the drab necessity of making a living. He was an avid botanist as a youngster. He lost his enthusiasm in college (Oxford, B.A. 1876), and spent three years studying engineering. Somehow, he became turned on again and went to Germany to study with **Sachs,** working with "milk vessels" in plants (Ph.D. 1882). There followed several years of miscellaneous lecturing—he was said to have a casual "hearth rug" style in addressing scientific meetings—and research in the Jodrell Laboratory at Kew on a variety of anatomical problems. Subsequently, he was honorary director of the Jodrell Laboratory for some years and was freed "from the daily grind of academic duties." Fourteen years later he "retired to the country" (some get all of the breaks, don't they). Now he was able better to pursue his professional avocation, paleobotany, and classification of the plant kingdom. The paleobotany came about as a consequence of friendship with W.C. Williamson, then the elderly dean of British paleobotany, who wished to turn his fossil collection over to someone who would maintain his legacy. The young collaborator did so.

Scott married in 1887. It is said that he had a happy domestic existence, but it certainly was marred by tragedy. Two children lived beyond infancy, and both died young, one of them in World War I. Scott's wife, Henderina, was a gifted artist who illustrated some of his books. She (as Rina Scott) also devised a technique for time-lapse photography that was "one of the earliest, if not the earliest, to attempt its use in plant physiology."

A broadly scholarly individual, who liked **Aristotle** but could not understand Plato, Scott possessed a gracious, informal mien. He was a charming writer. But this cloaked a stern, disciplined seeker of truth. One author has compared him with **Darwin,** especially with respect to his power of dwelling on one subject for years in seeking an objective solution. Perhaps so, but such similarity could perhaps more closely relate to the fact that both were of independent means.

Scott wrote a couple of "admirable" textbooks (1896), but he is best known to posterity for his work in paleobotany, where his accomplishments were second to none. These include descriptions and interpretation of numerous fossil genera, particularly of the Carboniferous flora, and perhaps more importantly, syntheses of the state of the art in books. His *Studies in Fossil Botany* first came out in 1900; it went through three editions. A book written in his senior years was *Extinct Plants and Problems of Evolution* (1922).

Of his work otherwise, I give but one example, a jigsaw puzzle in which he played *a,* if not *the* major role. I hope I am not confused in trying to reconstruct the following: *Lyginodendron* consisted of a fossil stem, variously referred to as the Cycadaceae or Osmundaceae. *Sphenopteris* was the foliage of a "fern." *Rachiopteris* was represented by a petiole. *Lagenostoma* included a cycad-like seed. *Crossotheca* was a fern-like, microsporangium-bearing male frond. These, he demonstrated, were all the same thing, a seed-bearing plant of fern-like aspect. Thus the *Pteridosperms*—phanerogamous "ferns"! An unknown proportion of presumed ferns of the Carboniferous flora possibly were these things, it seemed. About the Pteridosperms, Scott and others dealt with numerous questions,

Dukinfield Henry Scott

for example, a group of parallel development with the ferns? Another origin of the seed habit? Related to the Gymnosperms, especially the Cycads?

Paleobotanists have now gone well beyond Scott, but I am not sure that all of this has yet been satisfactorily resolved. That is beside the point. Scott materially broadened both the horizons of paleobotany and classification of the plant kingdom.

SELECTED BIBLIOGRAPHY

Arber, A. 1934. Dukinfeld Henry Scott. Nature 174:992-993.
Bower, F.O. 1938. D.H. Scott. Sixty Years of Botany in Britain (1875-1935), Macmillan, London, pp. 66-68.
Wardlow, C.W. 1975. Scott. In: Dictionary of Scientific Biography. Vol. 12. pp. 258-260.

Frederick Orpen Bower (1855–1948)

Bower was center stage during the tumultuous time (ca. 1880-1920) when relationships between the major groups of plants were perhaps most actively debated. Although Bower enunciated several theses, some certainly equivocal, he qualified these by the statement, "A working hypothesis, open to refutation, is better than no hypothesis at all." And he was never hypothesis-unarmed.

Bower was born in Ripon, England, of a well-to-do family. Little science was available in his formal education, Trinity College, Cambridge (A.B. 1877), and he bewails this in autobiographical writing (Bower 1938). His botanical education and inspiration came from association with a who's who of the major continental botanists of the time, for example, **Sachs** and **de Bary.** Communication with these luminaries followed by inspired research gave him a reputation that resulted in his appointment as Regius Professor of Botany at Glasgow in 1885 at the age of twenty-nine. There he stayed writing voluminously until and subsequent to formal retirement in 1925. Some of his most important books came out in the latter period.

I find little about Bower's human relationships. His interests, evidently, were entirely his work and, at least early in life, music. He played the cello and writes almost lyrically of the concerts of the

Cambridge University Musical Society in which he participated (Bower 1938, pp. 17-19). Possibly Bower was a bit of a snob. Brimble (1948) says that he "never wasted his time with any botanist whose work he suspected as not being first-rate." Walton (1948) says that "those who got beyond this rather formidable facade found a man who was very charming." It is agreed that he was an inspiring lecturer.

Bower's teaching activities are reflected in several textbooks and a translation (with **D.H. Scott**) of de Bary's *Comparative Anatomy*.

The focus of his most important life-long interest is reflected in the titles of two books, the *Origin of the Land Flora* (1908) and its successor *Primitive Land Plants* (1935). If you feel shortchanged by just the titles and brief commentary following, look at the books yourself. They are fascinating and can be found in many libraries.

A massive three-volume opus, *The Ferns,* was one in which pteridophyte classification, it is said, was changed from "convenience to phylogeny." The book *Size and Form* concerned the thesis that as plants get bigger, a certain ratio of size to bulk must be maintained. The devices are various. Bower's semiautobiographical *Sixty Years of Botany,* finished when he was eighty-three, is a kaleidoscope of the botanical period in which he worked and the people who inhabited it.

Obviously Bower was both brilliant and industrious. Three of the problems, among others, he tackled were the following: (1) Land plants are descending from aquatic ancestors. How was the crawling out of the water onto the virgin soil accomplished, and what did it mean in terms of reproductive necessities? (2) Is the Bryophyte sporophyte equivalent to the diploid phase of algae (the homologous theory), or is it a *de novo* innovation in plant evolution (the antithetic theory)? (3) Then the appalling gap between the Bryophytes and Pteridophytes: In the former, it is the gametophyte that is the green plant; the sporophyte is a brief incident on the way to producing more gametophytes. But this is reversed in the ferns and their "allies." Here the sporophyte is "the plant," one often capable of maintaining itself indefinitely through vegetative growth. The gametophytes are still there; they bear archegonia and antheridia which are obviously(?) homologous with those of the

Bryophytes. But these are inconspicuous elements of the life cycle. How did this transformation come about, or was there a transformation? Do the Bryophytes represent a line of derivation independent of that of the ferns and just a dead end?

Hofmeister was the one who saw through this muddle. This done, Bower vigorously espoused the antithetic presumption. And by the time of his *Primitive Land Plants* (1935), he was now armed with fossils of *Rhynia* and like things, and he and others saw these as the originals of the vascular plants. They had little to do with the Bryophytes.

A unanimous resolution of several of Bower's theses is yet to be achieved. The attention of most botanists is elsewhere. But much of what we teach about life cycles in beginning botany is derived directly or indirectly from Hofmeister and Bower.

SELECTED BIBLIOGRAPHY

Bower, F.O. 1938. Sixty Years of Botany in Britain (1875-1935). Macmillan and Co., London. 112 pp.
   With citations, p. 97; with sketches of the Dramatis Personae of his period.
Goodwin, H. 1970. Bower, Frederick Orpen. In: Dictionary of Scientific Biography. Vol. 2, pp. 370-372.
Green, J.R. 1909. Bower. In: A History of Botany 1860-1900, pp. 54-77.
   Role of Bower and others in developing the antithetic theory of alternation of generations.
Walton, J., E.J. Salisbury, L.J.F. Brimble. 1948. Obituaries: Prof. F.O. Bower. Nature 161:753-755.

Karl Ritter Eberhard von Goebel

(1855–1932)

Karl Goebel (as he abbreviated his name) was yet another of those Germans I knew of as an undergraduate. The physical weight and substance of his *Organographie der Pflanzen* frightened me and I left it alone.

Goebel's father was an industrial administrator. The son started education in theology at the wish of his mother, but was converted to science by lectures of **Hofmeister.** Then he went to Strassburg as an assistant to **de Bary** where he received his doctorate. Then to Würzburg with **Sachs.** A dazzling duo of mentors! Subsequent to a sequence of escalating appointments, he finally settled at Munich in 1891. There he moved the botanical laboratories from crowded quarters to a peripheral park where there was room for greenhouses and a garden. He initiated and developed both, not just for research, but as public showplaces.

Goebel was both a laboratory/closet and field botanist. His residence duties were constantly interrupted by lengthy trips to the tropics that continued almost to the end of his life. One wonders how he managed and financed these numerous leaves.

Nothing is said of wife or family in some three independent references, so one presumes Goebel was a bachelor. Remarks about his personality suggest that he was an admired and impressive figure. A sample: "He walked the earth with god-like serenity"; "tall and handsome personality"; his "imposing stature and massive head."

Unlike some of his German predecessors, Goebel was evidently a modest person. Having been made a knight by royal order, he said to his class that although a knight, he could and would still walk to lecture. At the time of his formal retirement, shortly before his death, he quit with the next to last lecture to avoid the inevitable student demonstration with the finale.

Goebel was said to be a forceful and dynamic lecturer but aloof in the lecture hall and laboratory. One writer suggests that the latter was a pedagogic device, inasmuch as he was genial in the field. That he was sometimes frustrated as a lecturer is suggested by the statement, "If an angel from heaven came down to give the botany lectures, the medics would still not turn up."

Beyond teaching, Goebel was editor of the journal *Flora* most of his life. He contributed several chapters to a revision of Sach's textbook. But it was Goebel's research and writing that not only supported his teaching but provided the basis for his reputation. The substance of much of his work and views is mostly included in his encyclopedic *Organographie der Pflazen* (1898-1901) which subsequently went through several editions and translations. I earlier pointed to his repertoire of mentors, but he was the protégé of none of them; he was his own man, and indeed in his *Organographie* he displays himself as a horse of a different color. Would he currently be classed as a morphologist, a physiologist, or a physiological ecologist? Perhaps the latter, or maybe reasonably listed as a monotypic genus unto himself. He steered clear of the megababble of his time.

"Morphology," he said, "is that which we do not understand physiologically." Physiologists have laid claim to him at least to the extent that the most informative review of Goebel that I have seen was published in the journal *Plant Physiology* (Lloyd 1935). Lloyd says that "all morphology must take account of or begin with his [Goebel's] organography." Then what is this organography? I fear I do not completely understand and limit myself to listing a few conspicuous features. Goebel, writing when taxonomy was in the era of its exuberant early Darwinian excitement, said that phylogenetic

Karl von Goebel

speculation based on comparative morphology is nuts. It is as futile, he averred, as picking up the confetti scattered about after a carnival at Munich and trying to return it to its original sacks. Perhaps such postures, certainly overstated, have turned some individuals away from Goebel. But the case is there: structure is commonly independent of phyletic relationships. Function determines structure, and structure determines function. Organs have an inherited level of flexibility to adjust structure to function. This can be determined by physiological experiments, for example, a xerophyte normally lacking leaves may produce these organs if placed under a moist environment.

But the variety of plant forms much exceeds the requirements of different growth conditions. Many of these are not adaptive; rather they are indifferent and drift with those of selective importance, generation to generation. Selection is not involved in their fate. Some fade away, but others may continue in diverse ways and emerge with adaptive value of a new kind.

Goebel rejected the view of Goethe that vascular plants have certain fundamental parts as stem and leaves and that all others are derived modifications. Rather he said that there are no organs that have an a priori existence; these are but human abstractions, for example, "In nature there are neither leaves nor lateral members."

Goebel's experimentation was often the simplest possible: a plant and a flower pot, and he was contemptuous of those who in effect hauled out a bulldozer to cultivate a flower bed. "There are those who think that because they can use a microtome that they are botanists," and "Why use a microtome when a hand section will do?" All right, but this posture has led Stafleu (1970) to comment that "experimentation took place only in his imagination."

Endowed with a broadly philosophical mind, Goebel was assuredly a maverick out of the main stream of emerging botany of the latter nineteenth and early twentieth century. Seemingly, he now has but few active proponents (two possibly somewhat of the same cut are Sherwin Carlquist of Rancho Santa Ana and perhaps more directly, Don Kaplan of Berkeley). Nonetheless many of his precepts have diffused into and affected thinking in both interpretational morphology, phyletic theory, and, I suspect, in physiological ecology. He was a man who saw the light from a different angle and one with the courage to be different. As such, his performance was superb.

SELECTED BIBLIOGRAPHY

James, W.O. 1972. In: Dictionary of Scientific Biography. Vol. 5, pp. 437-439.
Lloyd, F.E. 1935. Karl Ritter von Goebel. Plant Physiology 10:203-208.
Stafleu, F.A. 1970. Karl Goebel's Organography. Taxon 19:114-116.

Andreas Franz Wilhelm Schimper
(1856–1901)

As an undergraduate, I was much impressed by a tremendous book (maybe four inches thick) by Schimper entitled *Plant Geography on a Physiological Basis* (1903). It was a translation of a German work of 1898 and now a good candidate for the present series. So I went about finding something about Schimper. To my horror, I discovered there was not a Schimper, rather there were four of them, all related and all botanists. "My" Schimper, however, turns out to be the big one, the author of the *Plant Geography* book and subject of this essay.

Andreas Schimper, son of W.P. Schimper, a moss specialist, obtained his doctorate at Strassburg, studying botany and mineralogy. Thereafter he was an academic and world-traveling nomad. He worked successively with **de Bary, Sachs,** and **Strasburger.** He had a series of temporal appointments, including one at Johns Hopkins in the United States in 1880. He was at Bonn 1883-1888, possibly the longest period. He became professor of botany at Basel in 1898, almost at the end of his life. Wherever he was, his residence was continually interrupted by trips to various parts of the tropics in both hemispheres.

Did Schimper marry? References say neither yea or nay. The peripatetic nature of his existence suggests the latter. Sanders

(1975) says he was interested in "literature and the arts" and though solid in research, "was impulsive in thought and speech." He died at the age of forty-five, presumably from a combination of malaria contracted in the tropics and diabetes.

It is more fruitful to turn to Schimper's accomplishments as a botanist. His first work was histological, relating primarily to the growth of starch grains in chloroplasts and leucoplasts. In it he contradicted both the prevailing ideas of the nature of starch grain formation and their prior identification with living material of the cell.

But his term with this and related investigations was relatively short. When in 1881-1882 he visited Florida and the West Indies, his interests turned to the complexities of environmental adaptation and how this related to the distribution of plants. This was the theme for the remainder of his life except for some "medicopharmaceutical" books. Initially he was called an oecologist or autecologist. I guess he would be called a physiological ecologist today.

Groom (1903) says work like that of Schimper's had not been done before because the physiologists were too busy in their laboratories and systematists in their herbaria. Indeed Schimper did not trod a beaten path. In summary: He concentrated on water availability and the morphological and physiological adaptations of xerophyllous plants. Particularly he was interested in plants that exist in physiologically dry habitats, for example, halophytes, littoral vegetation (he devoted much time to mangroves), epiphytes, and alpine plants. He studied the various mechanisms through which such plants are adapted to their sites. He seemed to be especially fascinated by epiphytes, which he regarded as derived from terrestrial ancestors of their group. Genealogically, the plants had climbed trees in search of more light, but in so doing, without roots in the soil, were physiologically in the desert. Those that succeeded in the ascent necessarily evolved adaptive mechanisms.

Schimper was careful to note reasons for exceptions to some of his generalizations. Desert annuals have no special adaptations, as do the cacti. Yes there is adaptation; it is evasion. They grow only during those brief periods when water is available and exist otherwise only as seeds. Epiphytes that are not xerophytic also evade by being limited to sheltered, moist situations. Some plants can be grown in gardens under conditions entirely different from those to which they are limited in nature. The reason may be that cultural care frees such plants from the competition that would otherwise assail them in natural habitats other than those to which specifically adapted.

Plant ecology is commonly dated to **Warming's** *Plant Communities* in 1895, and that is reasonable. But Schimper (1898) is a pioneer here of nearly equal rank.

SELECTED BIBLIOGRAPHY

Groom, P. 1903. A.F.W. Schimper: An Appreciation. In: Fisher, W.R., P. Groom, and I.B. Balfour eds., Plant Geography upon a Physiological Basis, pp. x-xvii.
　　English translation of Schimper, *Pflanzengeographie auf Physiological Grunlage*, 1898.
Sanders, A.P.M. 1975. A.F.W. Schimper. In: Dictionary of Scientific Biography. Vol. 12, pp. 165-167. with citations.
Schenck, H. 1901. A.F.W. Schimper. Berichte Deutschen Botanichen Gesellschaft 19:Achtzehnte General-Versammlung, pp. 54-70.
　　In German. The Achtzehnte is paged separately from that of the volume per se; with citations.

Liberty Hyde Bailey (1858–1954)

"Pure" botanists tend to be romanticists. They will drive umpteen miles to see rare prairie plants but ignore those of the urban human environment they can touch outside of their front door. (Edgar Anderson said something like this some years ago.) Likewise, they eschew the plants of the farm together with hogs that consume them; these are regarded as the mundane fodder of agriculturalists. But the supreme romanticist of all, L.H. Bailey, remains the preeminent protagonist of everything from corn and onions to begonias. And Bailey was more than a plant scientist; he was a phenomenon of creation, among the most broadly gifted biologists produced by God and/or nature.

L.H. Bailey was born in Michigan of a farming family, his father a fundamentalist community leader. An early interest in natural history led him to Michigan State University, where he studied botany and horticulture (B.S. 1882). He then went to Harvard to serve as an assistant to **Asa Gray.** He married Annette Smith at about this time; there were two children, female, one of whom, Ethyl Zoe Bailey, was his companion and co-worker in later life. Bailey was already publishing and making a reputation for himself, for in 1885 he returned to his alma mater as professor of horticulture and landscape gardening. But this was not for long, because in 1888 he was called to Cornell University as botanist and horticulturist. By now he was also a rural sociologist, poet, and philosopher.

Liberty Hyde Bailey

In 1903-1904 he became dean of the College of Agriculture and director of the Agricultural Experiment Station. There, amid strenuous administrative years, he built agriculture and the supporting basic plant and animal sciences from a tenuous beginning to that "first with the most." For example, among the several departments he established, botany, entomology, floriculture, plant pathology, and plant breeding became among the elite in the United States. In the early decades of this century, professional education in plant and animal science was scarcely complete unless it included at least some time at Cornell.

Bailey's years as dean and director at Cornell were stormy because of his numerous innovations and continuous budgetary imbroglio with the legislature. Bailey always wanted more! He said that he allotted twenty-five years of life to getting his education, twenty-five years to pay his debt to society, and twenty-five years to do as he pleased. Consistent with this thesis, he submitted his retirement from Cornell in 1908; it was not accepted, and final termination of his official duties did not take place until 1913. His longevity, however, provided him with the latter quarter of a century he wanted. What he did was to establish a private institution, the Bailey Hortorium near the campus, doing research, a torrent of writing, and world traveling.

Bailey evidently had incredible energy; as a scientist and writer he worked rapidly, had a fantastic memory for details, and was intolerant of inefficiency and sham. Publicly he was possessed with the personality of a leader, becoming rapidly the president of every society or group to which he belonged. He was a superlative showman in public speech, a Billy Graham for plants and rural life. I heard him once at Cornell when he was already well into his eighties. Among current orators, his style perhaps most closely resembled that of Jesse Jackson. Even though his voice now cracked from time to time, it yet mostly came in bell-like, resonant tones, lacking the harshness that Jackson sometimes exhibits. Having once heard Bailey you never forgot him.

Enough of Bailey as an administrator and an unexcelled pulpit pounder of botanical faith. Let us now know him as a researcher, a compiler of encyclopedias, and a diversely directed multidisciplinary writer and philosopher.

Much of Bailey's initial publication was curiously divided into two worlds. It included: (1) innumerable papers on *Carex* (sedges) of which he was the late nineteenth-century authority; and (2) research and publication on crop production physiology and breeding of cultivated plants. But this was only the beginning; in total they number more than 700 including some 60 books. Titles of a few of the latter will provide some idea of the scope. *The Manual of Cultivated Plants* (1924; revised in 1949), though now outdated, is still used because there is no substitute; the *Cyclopedia of American Horticulture* (1900-1902, multivolume, Bailey the instigator, manager, and editor for the contributions of more than 1,000 authors); *Survival of the Unlike* (evolution essays); *Botany: an Elementary Text* (1900); *The State and the Farmer* (1908); *The Country Life Movement in the United States* (1911); *Wind and Weather* (1916, poems); *What is Democracy?* (1918); *Universal Service, the Hope of Humanity* (1918); and *The Harvest of the Year to Tiller of the Soil* (1927). Many of his books went through several or numerous editions, and some were reprinted after his death. And they were not just little pamphlets as some of the titles might suggest. Most were 200 to 400 pages; the second edition (1949) of *The Manual of Cultivated Plants* runs to 1,116 pages; that of the first edition of the *Cyclopedia* came to slightly over 2,000 pages. Incredible production! From this sample of book titles, it is evident that Bailey, besides being a botanist and horticulturist, was a rural sociologist with political overtones, a conservationist, a poet, and philosopher.

Theodore Roosevelt, then U.S. president, appointed Bailey as chairman of a Country Life Commission. I know nothing about this assignment except that certain recommendations of the commission resulted in the establishment of the Parcel Post Service. Odd progeny!

Subsequent to Bailey's nineteenth-century fling with *Carex*, most of his botanical publications were systematic revisions of genera that included economic plants: *Rubus, Cucurbita, Vitus, Brassica,* and the palms perhaps predominate. The *Rubus* effort, a monstrous undertaking, was less than entirely successful because Bailey had no idea what was going on cytologically.

After retirement, the major thrust became pioneer work in the palm family. Bailey traveled, together with his daughter, Ethyl Zoe Bailey, to the wildernesses of all continents in search of palms. It is reported he said that he gave up traveling only when the daughter became too old to go with him. That, however, is not correct. His world junketing ceased only when, in 1949, he fell and broke his leg. In his pockets were airline tickets to various parts of tropical Africa.

Liberty Hyde Bailey lived an enviably full and accomplished life. He had extraordinary talents, and he used them. His impact, in the first half of this century, both in the development and popularization of botany and horticulture and in ideals for a more perfect society, was at least an order of magnitude beyond that of most of us. However, except for those who yet use his *Manual of Cultivated Plants,* I suppose he is now largely forgotten. But that does not mean he is not yet with us. The waters from the great rivers (prior to human pollution) lose their identity after passage into the ocean. But their refurbishing influence remains.

SELECTED BIBLIOGRAPHY

Lawrence, G.H.M. 1955. Liberty Hyde Bailey. Baileya 3:26-40, with citations.
McDaniels, L.H. 1975. Liberty Hyde Bailey as I knew him. Cornell Plantations 31(1):9-12.
Rodgers, A.D. 1970. Liberty Hyde Bailey. In: Dictionary of Scientific Biography. Vol. 1, pp. 395-397.
Rodgers, A.D. III. 1949. Liberty Hyde Bailey: A Story of American Plant Science. Princeton University. Press. 506 pp. Facsimile: Hafner Publ., N.Y., 1965,. with citations.
Winters, D.L. 1977. In: Dictionary of American Biography, Suppl. 5, pp. 30-32, with citations.

Roland Thaxter (1858–1932)

I took my only mycology course about 1936, and I still remember Roland Thaxter/Laboulbeniales.

Thaxter was born of evidently aristocratic New England stock. Both of his parents were said to be "active in literary studies"—his mother was a poet and his father knew more about Browning than anyone else. There is no mention of who made a living or how; presumably such a plebeian activity was unnecessary or at least not discussed.

Naturally young Thaxter went to Harvard, receiving his Ph.D. in 1888. His dissertation concerned the Entomophthoraceae, peculiar parasites of insects. He was a plant pathologist for three years with the Connecticut Experiment Station but returned to Valhalla (Harvard) as soon as possible and remained as mycologist the rest of his life.

Roland Thaxter married in 1887. There were four children. He traveled extensively in the tropics in connection with his research. He was president of the Botanical Society of America in 1909. He was a Unitarian who had a mania about the evils of tobacco. With stern loyalty to his work, he felt that true advances in culture derived from search for the truth rather than in its application. He was "tall, of great dignity, poise and self-restraint, austere," (but also with a "dry" sense of humor). He was a music and art buff and said to be an accomplished musician. He regarded popularization as "exploitation of science," and though much traveled, was contemptuous of travelogues. Is this pure Harvardian?

Well, then, what did Thaxter do? Almost it seems, it is unimportant what he did; it was how he did it. He is said to have had an unparalled knowledge of the fungi as a whole. He set new standards in precision of methodology, not only of microscopic skill and manual dexterity, but of illustrative clarity and sophisticated interpretation. The style of Thackery, whom he is said to have greatly admired, however, does not easily adapt to mycological description, and some botanists have had less than praise for his written text. The characteristics of his teaching are described by Weston (1933a).

During his short term as a plant pathologist, Thaxter worked on *Gymnosporangium* (apple rust), onion smut, potato scab, and devised new spraying equipment (Weston 1933b). On return to Harvard, he initiated continued studies of the Laboulbeniales (insect exoskeleton inhabitants) of a quality that made the words *Thaxter* and *Laboulbeniales* almost inseparable. His monograph (1896-1931) was five volumes worth, which he illustrated himself with plates "exquisite in their execution." (A rave review is provided by Weston (1933a); a wrap-up sixth volume was cancelled by his death.) He also worked on the Myxobacteria and miscellaneous other, usually anomalous, groups of fungi. I reiterate, it was less what he studied than how he did it. That was what rendered him Roland Thaxter. As a teacher he better flourished with graduate students than undergraduates. He was a stern disciplinarian who expected the same kind of devotion to work that he exemplified. Even though his official term as a pathologist was brief (Weston 1933b), he is said to have had a major influence on plant pathology through students who worked with him for a mycological degree but took up plant pathology as their profession.

W.H. Weston (1933a) in a memorial essay said he was the greatest mycologist of his time. Perhaps so.

SELECTED BIBLIOGRAPHY

Lamb, I.M. 1976. Thaxter. In: Dictionary of Scientific Biography. Vol. 13, pp. 299-300, with citations.
Weston, W.H. Jr. 1933a. Roland Thaxter. Mycologia 25:69-89, with citations.
Weston, W.H. Jr. 1933b. Roland Thaxter (1858-1932), his influence on plant pathology. Phytopathology 23:565-571.

Nathaniel Lord Britton (1859–1934)

Most of you who have taken a plant taxonomy course in the eastern half of the United States (and often westward) have heard of Britton through reference to the New Britton and Brown *Illustrated Flora,* three volumes of it. But that was not the real Britton and Brown you were using in class, I hope. It was also three volumes, two editions, published 1896-1898 and 1913. Though now obnoxiously dated and using a deviant nomenclature, it has been reprinted umpteen times since and in fact is still sold to unwary victims. The New Britton and Brown is now getting old enough. It was written by **H.A. Gleason** and published in 1952. Gleason's was entirely a new book, not another edition of the classic. Gleason would have done well to have used another title to avoid confusion.

Britton was both a prolific author and the man who built the New York Botanical Garden from genesis to become one of the world's major research centers for systematic botany. He also was a pain in several places to many of his professional peers.

N.L. Britton was born in New York City and received both his B.S. (1879) and his Ph.D. (1881) in geology and mining engineering at Columbia University. Botany got into the act during several years employment with the New Jersey Geological Survey, during which time he prepared a checklist of the plants of that state. He then taught at Columbia for several years. His assignments started with

Nathaniel Lord Britton

geology, but soon became botany; he became full professor of botany by 1891. He then left Columbia in 1896 to assume directorship of the newly established New York Botanical Garden and stayed with this job until retirement in 1929 at age seventy.

Britton was an unimposing scrawny little chap ("of slight and apparently frail physique"), who sported a scraggly beard that looked as if it had been subject to unsuccessful herbicide treatments. He married Elizabeth Knight in 1885. There were no children. Mrs. Britton was a bryologist, indeed one of the foremost in North America during her time.

The generative idea for the New York Botanical Garden came to Britton and his wife while he was still at Columbia University on occasion of a visit to the Kew Garden. "Why can't we have the same thing?"

Getting back to the United States, Britton put the idea into action. A committee was organized to consider the matter. Britton was a member of said committee. They had first to obtain the necessary real estate (in the north Bronx) and fiscal support from the city for the administrative/herbarium building and the promise of an annual support allotment. And who was to be the director? Britton, of course, when the garden actually came to light in 1906. In his administrative role, Britton was a successful entrepreneur, who rapidly built the research resources (personnel, library, laboratory, herbarium) of the garden to second to none in the country. And keep in mind that Britton was not just author and director of research; with the devoted assistance of **J.K. Small,** he was manager of the whole shebang, the spokesman for its fiscal growth, its public relations, its horticultural direction, and everything else that falls to an individual or his appointees responsible for a major public institution.

Britton rapidly finagled a floristic and revisionary research group among whom P.A. Rydberg and J.K. Small were the most predominant actors for three decades. The big things authored by these men were regional floras and contributions to a multivolume *North American Flora*. Examples of the former are Small's two editions (1903, 1933) of a *Manual of the Southeastern Flora* and Rydberg's *Flora of the Rocky Mountains* (1917, 1922) and *Flora of the Prairies and Plains* (1932).

The boss was an administrative autocrat who (verbal recounting) set daily production (number of descriptions) quotas for the research staff. And one could not try to catch up by working late because the lights were turned off at 5:00 P.M. (J.K. Small tried to beat this by taking home one specimen of each species he was describing—a possible explanation of some of the eccentricities in keys and descriptions in his floras.)

Britton's own publications fill a 30-page bibliography. Large icebergs include first the *Illustrated Flora* for the northeastern states mentioned previously. The "junior author," Brown, in case you were wondering, was Judge Addison Brown, a person who contributed money, $25,000, for the establishment of the garden. And there were numerous contributions by Britton and other authors to the *North American Flora*, which, though going to many volumes, was never completed. Britton also published voluminously on the flora of the West Indies. Among several monographic pen-children, the biggie was Britton and Rose, *The Cactaceae of the World*, four volumes (1919-1923; 124 genera, 1,237 specimens). I suspect Rose should have been the senior author.

But all of this was probably not the chief basis for Britton's notoriety in his day. An International Code for Botanical Nomenclature was first established in 1871 as a means of bringing some order from the chaos that then existed in plant nomenclature. And it was periodically revised and indeed was reasonably accepted worldwide. That is, until Britton came along; he would have none of it and staged a one person revolt. After gaining some disciples and amid many hot words, his American Code of Botanical Nomenclature came to documented life in 1892. The consequence was the coexistence of two nomenclatural languages in U.S. journals and floristic books for about half a century.

Harvard University resolutely stuck by the International Code in the various editions of the **Gray's** *Manual* of this period, and so did most of the other major American taxonomic research institutions. On the other hand, they were more than matched in volume by the floristic and revisionary outpouring from New York, and the American Code was also taken up by the U.S. Department of Agriculture. As long as Britton was there, the American Code persisted. But after his death, it rapidly died. Small's 1933 *Southeastern*

Flora was the last manual using it, and the final gasp was in the early 1940s when the USDA gave it up. Thus at long last, American botanists got back to speaking the same tongue, not only among themselves, but as the rest of the botanical world. All botanists suffered under the impact of this schism. Those who were not systematists either died laughing at the woolly-headiness of their herbarium brethren or groaned because the only stable name for the soybean was soybean.

Recovery from this nomenclatural chaos took some time. Even in the 1960s, Britton was Shinners' (1962) chief villain in a moderately scurrilous, perhaps libelous, paper about some of his floristic predecessors. Some sample quotes: "It was a period of great corruption in public life . . . Britton stooped his lowest. . . . The American Code was a Brittonian device for achieving political power." Of the *Illustrated Flora* "with its crude drawings, slovenly taxonomy and outrageous nomenclature . . . it was the most backward step taken in American botany . . . the hundreds of species with unfamiliar names . . . it was an act of imperialist aggression."

Activism may indeed bring counterreaction, but this was a bit overdone. The *Illustrated Flora* was really not that bad, and if one wants to be an imperialist, there are better ways of accomplishing it than writing a botany book. Indeed, it is necessary to give vigorous credit to Britton for many things, including his triple role as a compulsive author, a taxonomic and nomenclatural activist, and an administrator.

Beyond the various conventional honors that I rarely list for these people (such can be taken for granted), it can be seen that an unusual number of genera bear Britton's name: *Brittonamra Kuntze* (Leguminosae), *Brittonastrum Briquet* (Labiateae), *Brittonella Rusby* (Malphigiaceae), and another *Brittonella* (a genus of beetles; Shinners would have liked that). *Brittonrosea* R.S. Williams is for both Britton and Rose (Cactaceae). About seventy species and varieties in diverse groups also bear the Britton name. And, of course, the journal *Brittonia,* well known in American taxonomic botany.

One might love Britton (perhaps difficult) or hate him, but every botanist of whatever calling from 1890 to 1940 knew who he was. Posterity cannot ignore the man.

SELECTED BIBLIOGRAPHY

Howe, M.A. and C.H. Woodword. 1934. Britton. J. New York Bot. Garden 35:168-180.
Merrill, E.D. 1938. Nathaniel Lord Britton. National Acad. Sci. Biographical Memoirs 19:147-159.
Shinners, L.H. 1962. Evolution of the Gray's and Small's manual ranges. Sida 1:1-31.
Stafleu, F.A. and R.S. Cowan. 1976. Britton. In: Taxonomic Literature. Vol. 1, pp. 332-348, with citations.

George Washington Carver (1864–1943)

Carver is someone you probably have heard of. He and Martin Luther King, Jr. perhaps come as the most immediate names in the history of the welfare of African-Americans in this country. Granted they worked in entirely different ways. But even though Carver's principal tools were plants, you likely may wonder if we are slightly stretching the definition to list Carver as a botanist. But no! True, Carver was interdisciplinary. The sociologists have sometimes claimed him. He has also been said to have been a chemist and a pioneer in chemurgy. But let us remember that his degrees were in botany with Pammel at (then) Iowa State College and that his botanical inheritance derived from Pammel. (Carver wrote many years later that Pammel had been the greatest single influence in his life.) Like Pammel, Carver was primarily an applied botanist who somewhat professionally changed direction. But this, of course, is not unusual, and the botany of many has been variously oriented. Some regard it as a stepping stone to administration and soon are no longer botanists. But Carver remained a plant scientist to the end of his time.

Among our curious breed, only **Darwin** and perhaps **Luther Burbank** can compete with Carver in the number of books about him and his work. Carver has been adulated. But like Darwin, Carver has also been debunked, even vilified.

The only certain generalization about the first part of Carver's life is that it was rough. The accuracy of his scanty childhood memories is commonly mistrusted by historians who otherwise differ among themselves in speculation. The date of Carver's birth, given above, is thus an approximation and perhaps the remainder of this paragraph is the same. Carver was born in Diamond Grove (now Diamond) in southwest Missouri. His parents were slaves. There was an older brother, Jim, and (so Carver has said) two sisters. George's father was killed in an accident about the time of the son's birth. His mother, Mary, was owned by Moses and Susan Carver. Allegedly slave thieves kidnapped Mary Carver, son George, and possibly also his sisters (or had they died earlier?). Moses Carver managed to rescue George; the others were not heard of again. Jim subsequently died of smallpox. Thus at an early age George Carver was left with no blood relatives. And he was nearly dead with whooping cough when retrieved by the Carvers, who nursed him back to reasonable health and certainly saved his life. But the cough had injured his vocal cords and left him with an unnaturally high voice.

George lived with the Moses Carver family until adolescence when he began wandering from place to place looking for an opportunity for education. Via several years in Kansas where he did everything from laundry to a try at homesteading, he came to Winterset, Iowa, and there made friends who encouraged him to enroll in Simpson College in nearby Indianola. This took encouraging because he had previously tried a Kansas college and had been rejected on the basis of color (Kremer 1987). But he made the jump and successfully enrolled at Simpson in 1890 with the idea of studying art and music. However, by next year, the "calling" that absorbed the rest of his life became manifest. He decided to leave the humanities as avocations and transferred to Iowa State College, Ames, to study scientific agriculture. There he was fortunate in becoming a special protégé of Pammel, the botanist, and to realize two degrees (B.A. and M.S.) under his direction. Young George was active in student affairs, was masseur for the football team, and continued his interest in music and art. A couple of his paintings received honorable mention at a Chicago exposition. After finishing his M.S. he was appointed assistant botanist in the agricultural

George Washington Carver

experiment station (Pammel bore the title of botanist).

But Carver did not remain as a faculty member at Iowa State College. The calling that absorbed the rest of his life became manifest and he accepted a position as director of agriculture at the Tuskegee, Alabama, Research Institute headed by Booker T. Washington. There his real career began and continued the rest of his life. His work was directed especially toward improvement of the dietary and living conditions of the people of the South, both black and white. Carver was a bit of an Edison in applied innovation with crop species of the South. His special plants were the peanut and the sweet potato from which he developed a legendary list of food, medical, and industrial products. He said that these two plants, which flourish in the southeastern states, alone are capable of supplying a balanced human diet. But he did not limit himself here. The tomato, then called the love apple, was occasionally grown as an ornamental and said to be poisonous. I have previously read (but can't now find the reference) that he would take a tomato or two to the podium with him when making a speech. He would eat the tomatoes while talking. He also derived products from cotton, soybeans, and waste materials. He prepared a traveling "Jesup Wagon" that carried an exhibit of the products that came out of his laboratory. And so on, beyond the limits of space here. Carver was not only a scientist, but a one-man extension service and humanitarian.

But not all was sweetness and light. Although he had been liked and respected at Iowa State and sent off with much fanfare, it was somewhat the opposite when he came to Tuskegee. He got off to a bad start by requesting two rooms (one as a sleeping-living room, the other his laboratory and art studio), but his colleagues were all stacked two to a room. They looked on him as a Northern outsider and did not like his fervent zealotry and self-centered attention to his role in the universe. They were not pleased that he was interested in the poor whites as well as blacks. Thus he was somewhat ostracized. It has been asserted that he had lived with whites so long that he did not know how to get along with those of his own race. He wanted to do research, nothing else, but he was also manager of two farms, had a full teaching load, and resented the several committees on which he had to serve (so what is new?). He was so busy and forgetful that his meager salary checks would accumulate uncashed on his table. Although there was evidently mutual respect between him and Booker T. Washington, he was probably an administrative nuisance. He wrote numerous letters (we would call them memos) to Washington (Booker T., that is) pleading release from some of his duties and for some (or more?) fiscal support. Evidently he did not get the unshackled freedom of action he wanted.

Actually Carver was little known beyond his own backyard until he appeared before the Ways and Means Committee of the U.S. Congress in 1921. (By then Booker Washington was deceased and Carver himself was getting along in years.) I have read the minutes of that meeting. He was given ten minutes to state an advocacy for a protective tariff on the peanut. He talked about and exhibited products he had derived from this lowly plant. These included crushed cake for flours, meals, and breakfast foods; ground hulls for burnishing metals; peanuts with chocolate for candy, salted peanuts, peanut bars stuck together with a sweet potato goo, a food combination of peanut and sweet potato, peanut butter, peanut "hearts" (the germs?) for feeding pigeons, flavoring for ice cream, skins for dyes, a diabetic food, peanut milk, peanut curds, a cereal coffee, salad oil, facial cream, ink, etc. He skillfully answered some barbed questions. The committee members were much entertained and he drew laughter. He received repeated allotments of another ten minutes and then another and kept going for an hour. He ended with applause and compliments of the chairman. This notoriety abruptly pulled up the stage curtains and catapulted Carver into national attention.

So Carver, as the other peanut grower, Jimmy Carter, became famous. Soon he was touring the South giving lectures about his plants or anything else. Unlike some current "name" speakers he was not charging $25,000 a session. The depression of the 1930s may have augmented his reputation because he was an expert on subsistence living—how to make do on very little. For whatever combination of reasons, the legend of G.W. Carver continued to grow and grow as the years went by, perhaps beyond reasonable proportions. He was an epitome of the syllogism that anyone (in the United States that is) can rise to heights seen through the clouds if they properly strive "to succeed, not just survive" in modern TV

George Washington Carver

lingo. Buildings were named for him, for example, at Iowa State University and at Simpson College, possibly others. Statues were made of him (Iowa State has one). Henry Ford became his friend. Ford financed and dedicated the George Washington Carver Museum at Tuskegee, and the erection of a G.W. Carver cabin at Dearborn, Michigan. Carver and Edison were acquainted and the latter purportedly offered him a six-figure job, which he turned down. A commemorative stamp came out after his death in 1948. **Linnaeus** had been more lavishly "stamped" in his country (Sweden) but Carver beat him to the draw in being the subject, together with Booker T. Washington, of a 50-cent piece. A movie, *Life of George Washington Carver*, was made in Hollywood (1938). The George Washington Carver National Monument was established at Diamond, Missouri, and finally in 1977, posthumously, he was enshrined in the Hall of Fame for Great Americans. Carver's papers at Tuskegee include 130 boxes of letters and similar material; there are smaller archives at Iowa State University and at the Carver Monument in Diamond, Missouri.

But to every deification, there comes also a debilitating counterreaction. The Carver story was and is no exception, and the negative view is perhaps best seen in a resounding boo by Barry MacKintosh in a book called *George Washington Carver: The Making of a Myth*. And both the bases and the cultural climate for this position were substantial. They derive in part from the 1960s "liberal" intolerance towards the Uncle Tom facade, evident in the philosophy of Booker T. Washington and certainly accepted by Carver: The black race, so they said, had to accommodate themselves to the situation as they found it and raise themselves by their bootstraps. With the coming of the Martin Luther King era, this doctrine was plain heresy. Then also the combination of continued resentment on the part of some of Carver's erstwhile colleagues and his solicitousness to whites added gasoline to these chilling flames. The fact that Carver never married and had a rather high voice (his juvenile whooping cough mentioned earlier) naturally was meat for the kind of speculation that investigative journalists love and to which any public figure may be subject.

Perhaps the tale of Carver partly can be interpreted in the light of the man himself. He was deeply religious and a bit of a mystic;

he said he had visions and was fanatically convinced that God had appointed him to do great things, to be a savior of blacks. But this passion evidently at times made him an obnoxious bighead among his immediate associates. And they and others could and did readily ask, how much did the furor of peanut activism really help the economic welfare of the South, especially of the blacks? And there, lacking documentation, I also have to wonder. Let us assume that he did "invent" all of the peanut products claimed. But was this followed by commercial development and production so that these would be widely available for peoples everywhere? I see no evidence of this except for peanut butter and the fact that he was probably responsible for defeating the notion that the tomato (love apple) was poisonous. Maybe this was enough. But yet the query remains: To what extent did this affect the well and fair of the South? Maybe an economist could (or has) answered this question. I cannot.

Some of Carver's feelings about all of this emerge in his letters, which I have slightly sampled in reproduction (Kremer 1987). As many of his time (and how did they find the time?) he was a great correspondent; for example, he wrote to Pammel until the latter's death. I refer you especially to Kremer, who samples his correspondence and to McMurray (1981) if you want more.

Carver's attention to the King's English was but casual but this did not keep him from expressing himself both forcibly and sentimentally. He said, "No individual has any right to come into the world and go out of it without leaving behind him distinct and legitimate reasons for having passed through it." Indeed it would be a better world if more followed this precept.

SELECTED BIBLIOGRAPHY

Carver, G.W. 1981. Plants as modified by man. Iowa State J. Research 55:209-217.
    A reprinting of Carver's B.A. thesis, 1894, and a short listing of Carver biography.
Kremer, G.R. 1987. George Washington Carver in his Own Words. University of Missouri Press, Columbia. 208 pp.
McMurray, L.O. 1981. George Washington Carver: Scientist and Symbol. Oxford University Press, Oxford. 367 pp.

Hugh Neville Dixon (1864–1944)
(and Mrs. Mary P. Dixon)

Hugh Dixon (sometimes Henry in literature) is not to be confused with **Henry Horatio Dixon,** a physiologist. He was, per **Winona Welch** (an important American moss person), the "Dean of British Bryologists." I am not sure this in itself provides the portfolio for inclusion among the more renowned botanists, but it will do. His credentials are scarcely negligible. And they are of an amateur whose labor was entirely for love.

Dixon, English, scion of a landed family, attended Christ's College, Cambridge (B.A. 1883, M.A. 1886). There he seemingly received a classical training minus any botany. But informal contact with a botany professor was evidently sufficient to give him a permanent infatuation with mosses. Bryophytes, however, remained a subsidiary career. This because Dixon's real job was headmaster of a Congregational-sponsored school for the deaf, where he soon became the director. That was his occupation until he retired in 1914.

He married in 1890; one child died, and no more were forthcoming. Minus next generation responsibilities, Dixon and his wife Mary were partners for life both in the school management and the mosses. Their deaths in 1944 were within a few months of one another.

Botanical praise for Dixon has been both euphoric and unanimous. Be there such a thing as the perfect English gentleman, that apparently was he, tolerant, scholarly, engaging, and humorous. His collecting vasculum, as of 1938, was inscribed with the names and dates of fifty European collecting trips. A churchly activist, he preached and lectured, and his assignments included, for some years, the post of director of the London Missionary Society of the Congregational Church. He played hockey until age fifty. He wrote poems and published a couple of books of these. His home was lauded for its beauty and lovely gardens. During his life, the herbarium was housed at home; after his death his British material went to Kew, the foreign holdings to the Botany Department of the British Museum.

All of this is very nice, but what did he do? Enough! He started out with British and European mosses, a manual of the British kinds coming out in 1886, subsequent editions following in 1904 and 1924. Of the last edition Edwin Bartram (1944), another bryologist, said, "It is likely to remain the standard for years to come." Then he turned to the mosses of the world, especially those of the European continent, but also southern Asia and the islands of the South Pacific. He wrote the *Bryophytes of New Zealand* (a book). He reported in journal papers on mosses of Celebes, Borneo, Sumatra, Fiji, South Africa, India, Siam, Madagascar, Ceylon, etc. His understanding of the bryology of the latter parts of the world was apparently based entirely on specimens, because his field work was limited to continental Europe and the British Isles. His published bibliography (Bartram 1944) runs six finely printed pages that I estimate to include some 250 to 300 items. While these are mostly short contributions, they include, however, not only descriptions of hundreds of new species, but revisions of "difficult" groups. Dixon's taxonomic judgment and intuition was said to be supreme.

H.N. Dixon and his wife led a good life, both for themselves and for their professions.

SELECTED BIBLIOGRAPHY

Bartram, E.G. 1944. Henry [sic] Neville Dixon. The Bryologist 47:136-144.
Richards, P.W. 1943-1944. Hugh Neville Dixon. Proc. Linnaean Soc. London 156:202-205.
Welch, W.H. 1944. Memories of H.N. Dixon. The Bryologist 47:145-146.

Deam is in the foreground.

Charles Clemon Deam (1865-1953)

Deam's *Flora of Indiana* was published in 1940. Taxonomists then regarded it as the ideal state flora, one which could serve as a standard by which future efforts might be judged. And it is still praised, fifty years later.

"Plain Charlie Deam," as he liked to be called, was a character who can but be little reproduced in these few lines. He was born in Bluffton, Indiana, where he lived and died. He tried a couple of years at DePauw University but decided that college was not for him. He then worked at various jobs, including that of an assistant to a druggist. The drug business became his thing, and in some way he obtained the funds to buy a store.

He was successful! It would be just as entertaining, or perhaps more so, to write of Deam the druggist rather than the botanist, but that is not for here except for a couple of examples, such as Deam's Nerve and Bone Liniment (good for Man or Beast). He said about female fingernails, "She's all right; she don't have those red vulture claws." But he was selling the makings for the claws.

Medical prescriptions and female fingernails, however, were not the talisman; they proved instead to be the means to an end. For Deam turned to botany and the drugstore was his National Science Foundation grant for life. As a pharmacist, he was a sharp professional; as a botanist he started as a rank amateur who became a professional, though not in the conventional sense. Amateur or not, he was state forester for twenty years, 1909-1928, minus a four-year break, which was due to the fact he grossly insulted the state governor.

Charles Deam married Stella Mullin in 1893. They had two children; a girl grew to maturity. Stella Deam was a loving caretaker and research assistant for her iconoclastic husband and joint participant throughout his amateur-professional life.

How did Deam do the *Flora of Indiana?* He devised a "weed wagon," a converted Model-T Ford with sleeping accommodations for two as well as room for botanical paraphernalia. He collected plants from every township (1,016 of them) in Indiana. He assembled a botanical library of several thousand volumes so that he would know what it was all about. He was correspondence-incontinent, writing thousands (?) of letters to hundreds of individuals, including all of the floristic botanists of the eastern states. He stated that there should be a special hell for those who did not answer letters. He visited most of the eastern herbaria looking for validation of Indiana reports for all species he had not himself found. He first wrote a book on the trees of Indiana (1912, with reprints and revisions to 1953), then the shrubs (1924), the grasses (1929, illustrated by botanist Weatherwax), and finally the *Flora* (1940).

Deam's pungent verbal and written commentary on botany and botanists (and everything else) became legendary. Of the botany of his day, Deam said, "I got Sinnott's 3rd edition of his botany. The book is ok for training college profs, but not for the mill run student. It is still necessary to write a beginner's book and get the horse before the cart."

And why was this amateur *Flora of Indiana* rated so highly? This, though it has neither descriptions nor illustrations, is a big, heavy thing of 1,200-plus pages you would scarcely want to take into the field. It does have excellent keys; it has Indiana distributional maps for all species and especially observational and/or critical notes about most. All of these, of course, contribute to its value. But more important, I suspect, is the meticulous workmanship that can be more easily felt by one who uses the book. Every apple that fell from a twig was checked, and if Deam said this one was wormy, you quickly knew you could believe him and not need to verify it yourself. Deam's remark about his book was, "If it is too faultless the reader will know I never wrote it, so you see where I am, between the devil and the deep sea."

Perhaps in broad scope, Deam was a minor character on the plant science scene. But he was sufficiently loved in Indiana that there are two major biographies of him—not many botanists have rated that. He will be remembered both as a person and a botanist by any student of the United States flora.

SELECTED BIBLIOGRAPHY

DenUyl, D. 1954. Charles Clemon Deam. Castanea 19:109-121.
Kriebel, R.S. 1987. Plain Ol' Charlie Deam. Purdue University Press, West Lafayette, Ind. 183 pp.
Weatherwax, P. 1971. Charles Clemon Deam: Hoosier Botanist. Indiana Magazine of History 62:197-267.

Albert Spear Hitchcock (1865-1935)

The name A.S. Hitchcock remains synonymous with grasses, especially those of the New World. But Hitchcock was more than the grass emperor of his time; he was a cosmopolitan botanical organization man and was influential in healing the divorce between the American and the International Code of Botanical Nomenclature, which had existed since the latter nineteenth century. And please note, A.S. Hitchcock is sometimes confused with C. Leo Hitchcock (no relationship), another U.S. taxonomic botanist.

A.S. Hitchcock was born in Michigan to Albert and Alice Jennings and was subsequently adopted by J.S. Hitchcock and wife. The time and circumstances of the adoption are not given in references I have seen. Hitchcock came to Iowa State College (B.S. 1884, M.S. 1886) where **Bessey** and Osborn were his principal mentors; then moved to the University of Iowa (1886-1889) as an instructor in chemistry; then to the Missouri Botanical Garden (1889-1901); then Kansas State College as professor of botany (1892-1901); then and finally to the U.S. Department of Agriculture in Washington, D.C., where he stayed the rest of his life becoming *the* agrostologist for the New World.

Hitchcock was married in Ames, Iowa, to an Ames woman named Rania Belle Dailey. They had five children. Professionally Hitchcock was said to have been a "kindly, cheery" soul, always

Albert Spear Hitchcock

enthusiastic, and consistently helpful to numerous herbarium visitors. His hobbies have been tabulated as "collecting postage stamps, tennis, and mountain climbing" (Anonymous 1937). I rather doubt that Hitchcock had much time for any of these except that he undoubtedly climbed mountains in pursuit of grasses. And professionally, he certainly climbed to the top of his chosen spire, grasses, and was more diverse than most of us in ascending a number of secondary peaks.

Firstly, Hitchcock was a field botanist, his collections (to some 25,000 numbers) being primarily of grasses of North and Central America and the Antilles, but extending occasionally to South America and the Old World. **Agnes Chase** (1936) uses a full-page column in the journal *Science* just to list his expeditions, many of them several months in length. On one trip, he (Hitchcock, 1890) used an original, maybe unique method of plant collecting. It was in Florida in 1890. He made a special wheelbarrow for his botanical and camping outfit and walked, covering 242 miles in 24 days, collecting all of the way. I like this, but fear it would scarcely be practical in that state now.

Hitchcock was also a herbarium director, that of grasses in the (now) National Herbarium. Chase (1936), states that he built it to the "largest and by far the most complete in the world."

But most importantly, Hitchcock did not just accumulate vast knowledge; he wrote about what he knew. A prolific author, he published both revisions of poorly understood genera and regional treatments through much of the Americas extending elsewhere to Hawaii. He (with much assistance from Agnes Chase) prepared the *Manual of Grasses of the United States*. It was published in 1935, only barely before his death. Its immediate acceptance (at $1.75) was such that the first printing sold in a couple of months. Its usefulness to applied plant scientists as well as taxonomic botanists continues to the present in the form of a second edition (1951) by Agnes Chase.

While Hitchcock wrote primarily of grasses, he prepared several small botanical texts of which *Methods of Descriptive Botany* (1925) was possibly most important. And he published several papers on the philosophy of plant nomenclature and the advocacy of the type concept. Here both through publication and action as chairman of committees of the Botanical Society of America plus participation in successive International Botanical Congresses, he hammered at the type concept, approximately as first defined in the American Code of Botanical Nomenclature. He was influential in its adoption in the International Code and in otherwise healing the wounds between the International Code and the rebel American Code, which had caused a nomenclatural schism among American botanists and between them and the international taxonomic community since the latter nineteenth century. At the Amsterdam International Congress in 1935, Hitchcock represented the Botanical Society of America in advocating a type register and establishment of an International Congress committee given the responsibility of preparing a descriptive listing of the world's herbaria. These ideas were subsequently conjoined in *Index Herbariorum*, which was subsequently developed and now has gone through eight editions. Hitchcock was also a proponent, through several congresses, of an International Bureau of Plant Taxonomy. Such has also come about.

Joseph Dalton Hooker once wrote "life is short and books are long." For Hitchcock, life was short, even at seventy years of age; there was much he yet wanted to do. But he died of a heart attack, shipboard, returning to the United States from the Amsterdam conference. The grass genus *Hitchcockella* bears his name. Jepson, of the University of California, in a copy of his *Flora of California* presented to Hitchcock, said of him, "eager explorer, far-seeing botanist and wise promoter of scientific research in America."

SELECTED BIBLIOGRAPHY

Anonymous (probably T.S. Sprague). 1936. A.S. Hitchcock. Kew Bull. Misc. Inform. 1936:107-109.
Anonymous. 1937. A. S. Hitchcock. National Cyclopedia of American Biography, Vol. 26, pp. 41-42.
Chase, A. 1936. Albert Spear Hitchcock. Science 83:222-224.
Escalona, F. ca. 1980. Albert Spear Hitchcock. Unpublished manuscript. Iowa State University, Ames, 29 pp.
Hitchcock, A.S. 1890. Camping in Florida. The Industrialist, Kansas State College, Manhattan.
    Seen only as a citation; reported by Escalona, 1980.
Stafleu, F.A. and R.S. Cowan. 1979. A.S. Hitchcock. Botanical Literature, Vol. 2, pp. 212-217, with citations.

Daniel Trembly MacDougal (1865-1958)

MacDougal, an American botanist, though a man of several faces, is best remembered as the pioneer physiological ecologist of the desert environment.

MacDougal was an Indiana product and graduated from DePauw University in that state in 1890. He received an M.S. from Purdue in 1891 and continued there as an instructor until 1893. Then places and doings stumble over one another for the next fifteen years. Indeed I first thought there surely was a twin brother taking part of the action. He (or the other) went to the University of Minnesota in 1893, married the same year, and stayed there until 1899. Then he migrated to the New York Botanical Garden as laboratory manager and administrator, remaining until 1905. But he was also a U.S. Department of Agriculture plant collector in the Southwest (1891) and Idaho (1892). Three years later, 1895-1896, he was studying in Germany with the supermen of plant physiology, **Pfeffer** and **Sachs**. Obviously he was not around Purdue much after he left in 1893. But he received a Ph.D. from that institution in 1897. The explanation seems to be that the Purdue folks so admired his dissertation, something on the curvature of roots, mostly prepared in Germany, that they dismissed all other requirements. The degree was awarded in absentia. I guess the twin can be dismissed.

So, I now double back to MacDougal's departure from the New

York Botanical Garden. We find he went west in 1905 to assume directorship of a desert laboratory in Tucson, Arizona, financed by the Carnegie Institute of Washington.

Some explanation is in order. If a plant scientist, you have probably heard of the Carnegie Institute through its sponsorship of botanical work over the years, but what was it? It derived from Dale Carnegie, a benevolent manufacturer-innovator and multimultimillionaire. The Institute of Washington was just one of a caboodle of trusts Carnegie established. It included a plant biology section. Shortly after the turn of the century, a board was assigned to consider a unit for desert studies in the southwestern United States. MacDougal, already with much experience in the deserts, was on that board. The laboratory opened in 1903 with Dr. W.A. Cannon as resident investigator. MacDougal then came as director in 1905. And he stayed with Carnegie until retirement in 1928. But he did not really retire. Living in Carmel, California, where a second laboratory had been established, MacDougal continued to work and publish as a research associate most of the rest of his long life.

About MacDougal himself. Pictures suggest that he was a large, even burly man. Commentary about him is unanimous about his energy, enthusiasm, and general zest for life. Rather than being one of those night owls, he did most of his voluminous writing early in the morning before others yet stirred. He was a good camp cook and probably a casual administrator, i.e., he gave no orders, he made suggestions. Scanning some of his books, I can assert firsthand that he was an excellent writer.

MacDougal's accomplishments were all over the botanical scene. While yet at New York, he was already exploring the American deserts. He became excited about **Hugo de Vries'** mutation theory and edited a de Vries book drawn from a series of U.S. lectures. This was the first introduction of de Vries to the English reading public. MacDougal tried working with *Oenothera* (which had been de Vries' baby) himself; he also attempted to induce mutations by injections into plant ovaries. He published a major paper on the effects of light on plants. He was an ingenious gadgeteer, devising instrumentation that he used extensively after moving to Tucson.

All right, MacDougal was evidently a good man who wandered from place to place. But what specifically did he do that yet attracts attention?

Daniel MacDougal

Indeed! In 1908, after only two years residence in the West, there came the book *Botanical Features of North American Deserts,* 110 pages and superbly illustrated. But the title does not properly serve the book. It includes everything: geology, geography, physiology, ecology, the dominant flora, and the happy adaptive mechanisms of its plants. MacDougal recorded rainfall, temperatures of air and soil, relative humidity, etc., for this and subsequent works. He studied especially the various coadaptations that allow plants to flourish in the temperature- and water-hostile environment. If he needed an instrument to measure or otherwise quantitate something, he just made it from whatever he had at hand. His dendrograph, commonly mentioned by later authors, was devised to gauge changes in tree trunk diameter. It was described and illustrated by him in his booklet *The Growth of Trees* (1921). The first model (p. 10 in that treatise) looks jerry-built and gives the feeling it might strangle the tree—a later one is simpler. The contraption was sufficiently sensitive that it could record and graph daily cyclic fluctuations in trunk diameter. Its "skills" contrasted with his dendrometer that recorded seasonal growth. I cannot determine from what I have read whether MacDougal was premier in developing the idea of dendrochronology to assay climatic changes over a period of some thousands of years. Be this as it may, MacDougal's work and publication was a continuing mix of laboratory and field studies. The latter included extensive visits to North African deserts to compare dynamics with those of North America.

MacDougal was not just a desert man. He wrote several plant physiology texts, apparently the first American physiologist significantly to compete with the Germans. To attempt to sample the nature of his writing firsthand, I invaded the library and picked four of his books at random. Taking these chronologically, the first (1908) was the descriptive *Botanical Features of North American Deserts.* Next, *Hydration and Growth* (1920), 176 pages. As the title indicates, it deals primarily with water relationships; prior concepts about the phenomenon of growth are there put to sleep. Next, *The Growth in Trees* (1921), mentioned previously with respect to instrumentation.

And finally *The Green Leaf* (1930), written after MacDougal's "retirement." That it is a book directed to a popular audience can be seen from the chapter titles, e.g., "The Grass Blade"; "A visit to

Green Leaf Mills"; "Dwellers in Darkness"; "Leaf Products and "Human Population." The "Darkness Dwellers" tells about fungi and their like. The "Green Leaf Mills" "enlarges a leaf sufficiently (to 30 to 40 feet thick) so that one can crawl into it (through stomata I suppose) and explore. It tells also that a pine tree has half an acre of leaf surface exposed to the sun. "Leaf Products and Human Population" should be explanatory by title—a forerunner of present literature on the increasing struggle between the ever growing human horde and the capabilities of the world's photosynthetic factories to keep up with it. Altogether a fascinating book. Of its period, without illustrations except for a few cremated line drawings, it lacks the glamour of present productions with their colored pictures and artwork to draw in nonbotanical readers. But that does not detract from its merit, and it is too bad that almost no one of present date has heard of it.

MacDougal's contributions to our knowledge of deserts and plant physiology were supreme. He was possibly also unique in his time in being almost a complete botanist, a feat that has now become essentially impossible. Additionally, he was probably the first physiologist to draw American students away from translations of German texts. We all are just visitors to this world, ships that pass in the night. MacDougal's voyage deserves our homage.

SELECTED BIBLIOGRAPHY

Long, E.R. 1958. Daniel Trembly MacDougal. Yearbook American Philosophical Society 1958, pp. 131-135.
MacDougal, D.T. 1930. The Green Leaf: The Major Activities of Plants in Sunlight. Appleton, N.Y. 141 pp.
Moore, G.T. 1939. Daniel Trembly MacDougal. Plant Physiology 14:190-193.
    This and Shreve, cited following, were contributions to an issue of Plant Physiology dedicated to MacDougal.
Murneek, A.E. 1958. Daniel Trembly MacDougal. Plant Physiology 33:383-384.
Shreve, F. 1939. Daniel Trembly MacDougal. Plant Physiology 14:133-197.

Mary Agnes Chase (1869-1963)

Okay, grass people, attention. Here is Agnes Chase of your **Hitchcock** and Chase *Manual*.

Agnes Chase was born Agnes Meara (or Mera) in rural eastern Illinois. She was one of several children in a blue-collar family. After her father died when she was two years old, the family moved northward to Chicago where the surname was changed to Merrill. Agnes' formal education was limited to grammar school. She then worked at miscellaneous jobs, including that of a proofreader and typesetter for a magazine called the *School Herald*. At age nineteen, she married the editor, William Chase, who died a year later. There were no children.

Ms. Chase became acquainted with a couple of amateur botanists who interested her in the flora of northern Illinois and by 1897 she was collecting plants. One reference says that she proofread by night and collected plants in the day. One of her botanical stimulators, E. J. Hill, a high school teacher, minister, and bryologist, employed her to prepare illustrations for him. Her work came to the attention of C.P. Millspaugh of the Chicago Museum (1901) for whom she prepared illustrations for a couple of his publications. This work apparently brought her to the attention of the botanical community. In 1902 she pulled up stakes and moved to Washington, D.C., earning $720 per year as a U.S. Department of

Agriculture illustrator. In 1905 she became associated with the young grass man, A.S. Hitchcock. Here her professional evolution began. She was first Hitchcock's assistant and then collaborator in publication, for example, by 1910 she was junior author on a revision of North American *Panicum*, 396 pages. And then she became also her own person as given in her *First Book of Grasses* (121 pages) written for the neophyte (1922). Evidently it was used; subsequent editions followed (1937 and 1959). Hitchcock died in 1936 shortly following publication of his *Manual of Grasses of the United States*. It is probable that he could never have completed it had it not been for Chase's constant help.

Inevitably Agnes Chase became the grass person following Hitchcock and advanced to senior botanist. Somewhere along here, the grass herbarium and systematic work went to the Smithsonian Institution, ultimately (1947) becoming part of the Department of Botany. And there Chase stayed, two-plus decades after formal retirement (1939) as an unsalaried research associate. Her first major production during this time was the second edition of the *Grass Manual*, now *Hitchcock and Chase* (1951). And then came a verified card index, some 80,000 cards, for all (worldwide) published names of grasses. Picking up an effort started by some of her predecessors, she likely did most of the total work for its completion, and this probably occupied much of her time the latter part of her long life. In connection with the index, she made several trips to European herbaria to locate and verify type specimens. This unparalleled reference for agrostologists was published (Chase and Niles 1962) and is available in most major herbaria.

From this one could get the impression that Chase was entirely a herbarium and bibliographic pedant, a "closet" botanist. That is not correct. She invaded the field in numerous trips until the age of seventy, not only in the United States but in a succession of expeditions to Mexico, the Antilles, and South America. Her collection accessions, mostly grasses, ran to about 12,000 numbers.

With this background, I came into this essay with the impression that Chase was probably a mousey woman who lived entirely for Poaceae (the grass family) and was a loner with limited social life. Mostly wrong. It is said that she was an excellent cook and loved to entertain guests. Once when Dr. Floyd McClure (bamboos) was

Mary Agnes Chase

with her for dinner, she accidentally burnt the cookies. Politely, he said no matter, he loved burnt cookies. But subsequently, when he was her guest with others, she made two pans of cookies, one of usual baking level, the other charred just for him. She was a woman suffragist who marched in parades advocating voting rights for women and was twice jailed. She was a prohibitionist and anti-smoking. Originally a Catholic, she abandoned the Church, and Stieber (1980) says that her christianity was socialism (alternatively, she might have made a good Unitarian). At different periods she shared her home with at least two other women. Among numerous stories about her, one relates that she would ask botanical visitors to the herbarium (there were many over the years) what group of plants they were interested in. If these were not of the Poaceae, she would then just walk away. If their answer was that holy word, grasses, she was their attentive helper the duration of their visit. In some instances, she provided them with lodging in her home during their stay.

At the time of her demise, Chase was perhaps the most important grass person worldwide. And she is credited by some as being the one who put the name agrostology on the botanical scene.

SELECTED BIBLIOGRAPHY

Chase, A. and C.D. Niles. 1962. Index to Grass Species. 3 vols. C.K. Hall and Co., Boston.
   This is not something I expect you to look up.
Stieber, M.T. 1980. Chase, Agnes Mary. In: Notable American Women: The Modern Period, p. 146-148. Belknap Press, Harvard University, Cambridge, Mass.
Swallen, J.R. 1959. Biographical sketch of Agnes Chase. Taxon 8:146-151.
   Includes Chase bibliography.

Henry Chandler Cowles (1869-1939)

The inspiration of **Warming** in Denmark came to the United States in the form of Cowles and **Clements** who received their Ph.D.s in the same year (1898). Although later overshadowed by Clements, Cowles beat him to the draw in writing about succession (1899) and perhaps became the most universally respected New World ecologist of his era.

Cowles was born in Connecticut of a farming family. He graduated from Oberlin College in 1893. He then spent a year teaching at Gates College in Nebraska. Then, when he went to the University of Chicago to study geology, he was captured by **John Merle Coulter,** botanist. Cowles remained at Chicago, a botanist, the rest of his life: graduate student, professor, and department chairman. After a youthful flush of personal research (ca. 1895-1902) and subsequent text writing (1910), his continuing essence was primarily that of his students. Of allegedly affable personality, he then spent most of his life as a teacher, an activist in professional societies, an editor, and departmental administrator.

Of Cowles' personal life, one reads only that he had a wife and daughter.

Cowles was an excellent writer. He produced two significant, even magnificent publications, both seemingly of parochial nature (i.e., limited to the Lake Michigan-Chicago environment) but from which, however, he boldly declaimed broad hypotheses (enunciated as principles) about the dynamics and classification of vegetation. The first was the sand dunes paper (1899; his doctoral thesis) in which the idea of succession leading to an ultimate equilibrium was less than cautiously expressed. The goal was to "arrange plant societies in order of development . . . the primitive plant societies pass rapidly or slowly into others . . . the plant assemblage more and more approaches the climax type of region." The second one, *The Physiographic Ecology of Chicago and Vicinity,* was ostentatiously subtitled *A Study of the Origin, Development and Classification of Plant Societies.* The objective was a "classification of plant societies which shall form a logical and connected whole." No false modesty here.

I do not know the extent to which Cowles' theses were really of his own genesis. He paid due homage to Warming and perhaps was mostly trying Warming out in his own backyard. In any event, the content of these papers, plus authorship of the Ecology volume of the then widely used Chicago botany text (1910-1911), firmly identified him as perhaps the first real American ecologist. Having thus achieved such status, he essentially rested on his personal research laurels, continuing his career primarily as related above, but intermittently asserting his authority as puritanical monitor of the morality of ecology. For example, he desired completely to eliminate any implications of teleology in ecological writing. One should say "response" not "adaptation," likewise, "accumulation of food reserves," not their "storage." "Evolutionary strategy," an expression common in current literature, would probably have given him a stroke.

Although Cowles was not the first to discuss succession, **Tansley** has stated that his was the "first working out a strikingly complete and beautiful successional series." That may be a good evaluation. And where he pioneered, successive human seres followed.

SELECTED BIBLIOGRAPHY

Cooper, W.S. 1935. Henry Chandler Cowles. Ecology 16:281-283.
Fuller, G.D. 1939. Henry Chandler Cowles. Science 90:362-364.
Kraus, E.J. 1939. Henry Chandler Cowles. Bot. Gaz. 101:241-242.
Sears, P.B. 1958. H.C. Cowles. In: Dictionary of American Biography, Suppl. 2, pp. 127-128, with citations.

Henry Horatio Dixon (1869–1953)

In the early nineteenth century, **Thomas Knight,** using a dye, observed the upward movement of sap in plants. He speculated about the forces that lifted (pulled?, pushed?) it upwards. Subsequent physiologists contributed theories, some reasonable, some less so, but none generally accepted. Conjectures remained contentious for nearly a century. Dixon, here, is eulogized by physiologists because his "cohesion theory," supported by both direct and indirect theoretical evidence, provided the basis for the present concepts about this vital plant process. Such lacking, our situation might resemble that of animal physiologists prior to Harvey's deductions about blood circulation.

Dixon (Irish, sometimes confounded bibliographically with **Hugh Neville Dixon,** a bryologist) came of a well-to-do family. The finale of some nine children, he entered Trinity College, Dublin (1887), with scholarships to study Italian. Possibly lacking in sufficient south European genes, he graduated a "classical scholar" specializing in natural history. He then went to Germany to study with **Strasburger.** But he was a homing pigeon, for after two years he returned to Dublin as botanical assistant, where he advanced to professor in 1904 and so remained until retirement in 1949.

Dixon married in 1907, and the family subsequently included three sons.

Most of Dixon's career activities were as you or I if we contin-

Henry Horatio Dixon

ued with the brute energy of youth. He taught school, including, among other courses, botany for medical students and general botany, which emerged as a text, *Practical Plant Biology*. He busied himself with miscellaneous minor research. He obtained funding for new and improved quarters for the botany "department," i.e., the professor plus an assistant(s?) and demonstrator. He expanded the botanical garden. He personally cared for the herbarium (containing W.H. Harvey algal types) and indeed negotiated a new wing on the building to better house it. All-fine; a good all-around botanist, but many others were doing most of the same things.

Except how water gets up a tall tree. Two circumstances facilitated achievement of "the most brilliant and original contribution to this central problem of plant physiology" (Morton 1981). First, Dixon had been with Strasburger when the latter was working on the anatomy of the phloem and xylem and searching for a relationship between structure and function. By killing the conducting cells with steam heat of trees to 30-feet tall, Strasburger found that upward water transport was scarcely affected; hence the process was evidently mechanical and not dependent on living cells. Secondly, Dixon had a companion at Dublin, physicist John Joly, who took him on yachting cruises. The two also got together on the water transport problem and the Dixon and Joly *On the Ascent of Sap* was published in 1894. Then in subsequent papers (1909, 1914, 1924) Dixon further polished the theoretical reasoning concerned. This, the cohesion or tension-cohesion theory, is the basis of present interpretations. I have not the space here to explain beyond an analogy, namely that the tensile strength of a fine column of water constitutes a thread pulled upwards by transpiration.

True, Dixon had the collaboration of Joly. And also a chap, Askenasy, came up with somewhat the same idea only a year after (1895) the Dixon/Joly paper. But we must acknowledge Dixon as the initial spirit who then nailed down the subject by his continued attention over a period of thirty years. On occasion, ordinary botanists succeed in extraordinary achievements.

SELECTED BIBLIOGRAPHY

Anonymous. 1939. Henry Horatio Dixon. Plant Physiology 14:616-619.

Harvey-Gibson, R.J. 1919. H.H. Dixon. In: Outlines of the History of Botany, pp. 192-194. Black Ltd., London.
Morton, A.G. 1981. H.H. Dixon. In: History of Botanical Science, p. 428. Academic Press, London.
Simpkins, D.M. 1971. H.H. Dixon. In: Dictionary of Scientific Biography. Vol. 4, pp. 130-131.

John Kunkel Small (1869–1938)

I was with the Tennessee Valley Authority summers 1943-1946 working with the malaria control research group. This is not the place to explain the role of a botanist in malaria control or the relationships between the kinds and abundance of plants growing in or close to the water and malaria mosquito breeding. (But if you are curious, refer to Tennessee Valley Authority 1947.) I was given the job of preparing an identification manual for the plants growing in the fluctuation zones of the reservoirs. There was no reference herbarium until I assembled one, and all that I had to work from was the then relatively new (1933) Small's *Manual for the Southeastern Flora*. I grew to loathe that book and found in following years that I was but one of a chorus. Adolescent judgment! The Small has continued to be useful for much of the southeast United States in the last forty-five years because of the paucity of state floras in the region. Small was a brave man who single-handedly did his best to conquer the impossible.

J.K. Small, born in Harrisburg, Pennsylvania, went to Franklin and Marshall College in that state and obtained his doctorate at Columbia University in 1895. Subsequently for a few years, he was employed as chore boy for **N.L. Britton** of the Columbia faculty. Then when Britton moved to the Bronx to become the first director of the just established New York Botanical Garden, he took with him Small and the Columbia herbarium. Small continued the chore boy role, "Curator of the Herbarium, the Museum," and everything else. This was his job. Britton made it clear that any research and writing was for the back shelf, one to which Small might turn to after everything else was done. Obviously everything else was never done. It is a marvel that Small wrote anything at all except interoffice memoranda, requests for money, reports, and especially correspondence with the inquiring public ("What is this plant and what is it good for?—such requests are innumerable in a large city).

John K. Small married Elizabeth Wheeler while yet at Columbia; they had four children. I find nothing about his personal life save that his was a musical family. He played several instruments, and is said to have been an excellent flutist. He insisted that each youngster learn two instruments, and it is related that the family often spent early evenings concertizing together. Then everybody went to bed except poppa, who burned the night writing.

The first production came to overwhelming fruition, the *Flora of the Southeastern United States,* 1,370 pages, 3½ inches thick, 1903, and published by the author (Britton believed that the garden should not get fiscally involved with books, excepting of course, those written by Britton). The enormous Small book sounds impossible. He had only graduated from college in 1895. Further the feat of even learning something of several thousand kinds of plants from scanty herbarium records, from **Chapman's** ancient (1860, 1883) flora of the southeast states, and from minor prior publications is appalling. Success in the effort to get all of this on paper, including keys and descriptions, seems to require help from some divine source in knowledge and penmanship. That such help was lacking is suggested by the fact that in some ways it was a rather cruddy manual, or, to be more kind, yet a pioneering one. It was obviously a product of youthful enthusiasm, energy, and guts. Small, of course, knew it was less than ideal and cleaned up some of the mess with a second edition in 1913. These books were used.

But Small knew this was not yet enough. Although he had first gone to the southeast as early as 1891, he really knew little about the plants in the field and the nature of the regions they inhabited. He set out to correct this and was successful in obtaining sponsors for travel (he didn't drive a car), field work, and subsistence. With this help he was then, for a quarter of a century, in the southeast for nearly every year on short or extended explorations in Florida and east Texas. The result eventually was his 1933 *Manual for the Southeastern Flora,* the one that I used in my initial contact with southern Dixie land.

And I fussed about this final opus magnum. Why?

First, the manner in which it was produced caused sometimes blatant discrepancies between the keys and the descriptions. They were written at different times (so it is said) and did not necessarily agree. Then also Small could not work at night at the garden. The gaslights did not provide sufficient illumination and Britton forbade night habitation by the scientific staff. But Small had to work the wee hours were he ever to finish. So, each eve, he would take a bundle of specimens home with him, one for each species he was to describe that night (Wherry 1957). Obviously the resulting descriptions were often but blurred and sometimes misleading images of his plants.

Nomenclatural and taxonomic dicta made Small's book difficult to use. Adhering to the American Code of Nomenclature (a religion of Britton's), the manual appeared at a time when this code was dying of something resembling Parkinson's disease; it spoke a language foreign to most systematists. In addition, Small used almost entirely binomials—there were no such things as varieties or subspecies. In south Florida, he found many kinds of plants which he described as new, but had he had time to look further, he would have seen that they were just northernmost extensions of tropical kinds of the Antilles.

Enough or perhaps too much. This is why others fumed. But we must remember that Small trod a mine field where these others feared to venture. Only as this is written has personnel of the University of North Carolina attempted to assemble a taxonomic army to try to improve on what Small did. They have not yet succeeded.

SELECTED BIBLIOGRAPHY

Barnhart, J.H. 1938. John Kunkel Small. Science 87:129-131.

Barnhart, J.H. 1938. The passing of Dr. Small. J. New York Bot. Gard. 39:73-79.
Tennessee Valley Authority. 1947. Malaria Control on Impounded Water. U.S. Government Printing Office, Washington, D.C. 422 pp.
Wherry, E.T. 1957. Reminiscences of John K. Small. Castanea 22:126-129.

Arthur George Tansley (1871-1955)

I suppose Tansley, British ecologist and approximately contemporary with **Clements,** was somewhat the English equivalent in his time. But Tansley was more diversely oriented than Clements and less insistent (or rigid) in his schemes of the classification of vegetation. Hence he has been less mangled by succeeding generations. Is that the reason why we in the United States yet hear much more of Clements than of Tansley? Or is it because we are parochial? Really the gregarious Tansley was a more interesting person than Clements.

Tansley, born in London, went to University College there, where he became a protégé of the paleobotanist-everything, F.W. Oliver. Then he transferred to Trinity College, Cambridge, then back to University College (1893-1906) as a member of the faculty, finally yo-yoing back to Cambridge in 1906. His forte was ecology and classification of the vegetation of the British Islands, and he was already illustrious in his profession. He was set for life.

But it didn't work out that way because in 1923, now middle-aged, he jumped the bandwagon and resigned.

He had become a Freudian. Already having written a book *The New Psychology and its Relation to Life* (1920), he now went to Vienna to study with Freud. Psychology, however, turned out to be but a piece of the pie, for he also continued botanical writing. In 1927, he formally returned to plants as Sherardian Professor of Botany at Oxford and stayed there until retirement in 1937. But he by no means retired as we shall see.

Arthur Tansley married one of his students in 1903 (like Clements' wife, her name was Edith); they had three daughters. As a teacher, graduate or undergraduate, he is unanimously said to have been an inspiration. Some professors are easily forgotten. He was not. Although an intense botanist and an activist for this and that all of his life, "on relaxed occasions, he was jovial and humorous" (Hope-Simpson 1971). He was tolerant of the foibles of others.

As an ecologist, Tansley was neither especially quantitative nor experimental. Rather he observed and let his intuitive syntheses do the rest. His primary interests, community structure and distribution of vegetation, were approximately analogous to those of Clements. But he was much more of an organization man than the commonly hooded Clements. He founded a British vegetation committee that published *Types of British Vegetation* (1911) of which he was the primary author. The British Ecological Society came into existence in 1913 with Tansley a leading sponsor, the first president, and subsequently editor for many years of its *Journal of Ecology* (Don't confuse with *Ecology* in the U.S.). Tansley was then doubly an editor because he had earlier founded *The New Phytologist* and remained at its helm some thirty years. Among his several books, the reference masterpiece was probably *The British Islands and Their Vegetation* (1939). But he was not just a famed botanist, his diversity being somewhat indicated by the following titles: *The Value of Science to Humanity* (1942), *Our Heritage of Wild Nature* (1945), and *Mind and Life* (1953).

One British spin-off of (or reaction to) the destruction of World War II was a program for the conservation of nature in which Tansley played a leading role. This brought him public attention and perhaps was the basis of his knighting in 1950. The British Nature Conservancy in part owes its founding to Tansley, and he became its first chairman, 1949-1953 at age seventy-seven. (Explanation for readers: the Nature Conservancy in England has nothing to do with that of the same name in the United States. It is a government agency, that in the U.S. is a private organization.)

The following excerpt from a letter (1950) written by Tansley published in the *London Times*, slightly brings him to life and is as cogent now as it was then:

> No form of higher education is, or can be, completely self-supporting. Undergraduates do not, and could not, pay the stipends of the academic staffs of universities. . . . The close contact with nature and her processes and their interaction with human activities, which practical field studies can give, is the only kind of training that builds up the necessary knowledge and mental attitude which cannot be created by book work or laboratory work alone. The council has made a start—but only a start. A.G. Tansley, President.

In tribute: Sir Arthur Tansley—a keen personable botanist who not only achieved international fame within his profession but went well beyond. A jolly fine fellow, what?

SELECTED BIBLIOGRAPHY

Anonymous. 1955. Sir Arthur Tansley. The London Times, Nov. 28, 1955, p. 13, column a.
Anonymous. 1956. Sir Arthur Tansley. The New Phytologist 55:145-146.
Hope-Simpson, J.F. 1971. Tansley. In: The Dictionary of National Biography (British). Supplement 8:953-954. Oxford University Press.

Richard Willstätter (1872-1942)

To botanists, Willstätter is primarily known as the man who first worked out the chemical composition of chlorophyll. But to Willstätter, an organic chemist, the chlorophyll story was but one of a sequence of scientific adventures.

German and Jewish, Willstätter received his doctorate at Munich in 1894. There followed a series of positions of increasing substance, and he came in 1912 to the post of director of the Kaiser Wilhelm Institute of Chemistry, Berlin-Dahlem. But he was less than happy in Berlin and in 1916 returned to Munich as professor of chemistry. World War I was then shaking the world, and Willstätter, his research much interrupted, teamed with Haber in designing gas masks for the German army. (These were subsequently duplicated by the Allies, following analysis of captured masks.) In odd contrast, in 1915 he received the Nobel prize for chemistry, this being based (scarcely on gas masks) primarily upon his prior work on chlorophyll.

Anti-Semitism blighted Willstätter's life and career. He was exposed to it even as a youngster. He resigned his post at Munich in 1924 in protest against the rejection of appointment of V.M. Goldschmidt, said to be the founder of geochemistry, to the chemistry faculty because of racial bias. Not only did he resign, but he refused other professional offers. No one explains how Willstätter

Richard Willstätter

supported himself during the ensuing years, but he apparently lived reasonably well and indirectly continued some of his research through Margarete Rohdewald of the Munich staff. The coming of the Hitler days further rendered life for professional Jews impossible. Willstätter escaped the fate of many of his race by fleeing to Switzerland in 1939, where he died three years later.

Willstätter was a small-statured, proper, and dignified chemist and university administrator. He married early in life; there were two children; then his wife died (1908); his son died at the age of ten. He did not marry again.

The sequence of some of Willstätter's major investigations, all botanical, was somewhat as follows. His doctorate concerned the structure of cocaine, and he continued for several years to work with related tropine alkaloids. Then he turned to the solanaceous alkaloids. Then, perhaps his masterpiece, chlorophyll, the nature and composition of which was a subject of controversy. In succinct summary, what he and his students found was close to what we yet teach in beginning botany. They said that chlorophyll of higher plants (ca. 200 species examined) includes a "mixture" of four compounds. These include two chlorophylls, chlorophyll-a and -b, with magnesium at the metal attachment, not iron as previously hypothesized. And there are two yellow to red pigments, the carotins and xanthophylls. The algae have these same pigments but are more diverse than the flowering plants in their chlorophylls and accessory pigments. Because the Willstätter group only moderately sampled the algae, they missed most of the variants, which were gradually filled in years later. But they provided the big picture, including the empirical formula for each and the approximate proportion that it made of the whole.

Then Willstätter moved to anthocyanin pigments of flowers . . . and then to enzyme chemistry. Indeed it seems to be a characteristic of Willstätter's work that while he pioneered in opening up new territories, he usually left further exploration to others; for example, he did not attempt the structural formula of the chlorophylls and made only a brief pass at their functioning. This continued saltation from one area of inquiry to another was perhaps due in part to his institutional moves and interruptions in his work. Or possibly it derived from his personality; after discovering new land, he lost interest and left exploration and settlement to others.

Richard Willstätter bore the triple cross of unusual gifts, the scars of two wars, and of racial intolerance. Through it all, seemingly free of arrogance and envy, he remained the perfect gentleman.

SELECTED BIBLIOGRAPHY

Fruton, J.S. 1976. Willstätter. In: Dictionary of Scientific Biography. Vol. 14, pp. 411-412.
Huisgen, R. 1961. Richard Willstätter. J. Chemical Education 38:10-15.
Robinson, R. 1953. Richard Willstätter. J. Chemical Society London. 1953:999-1026.

Merritt Lyndon Fernald (1873-1950)

Fifty years ago, the name Fernald was familiar to nearly every botanist in the United States. He is yet known to many of us as the author of that standard plant identification reference, the eighth edition of *Gray's Manual*. Others think of him in terms of provocative theories in plant geography. Although no one could dominate North American botany as once did Asa Gray, Fernald came close for systematic botany for some forty years, at least in eastern North America.

Dr. (honorary) Fernald was born in Orono, Maine. There his father was professor of physics and mathematics and later president of the institution that is now the University of Maine. At the age of seventeen, the younger Fernald published a couple of papers on *Carex*, followed by a checklist of plants collected near Orono; several were new to the state. There was no doubt about his profession to be. And it started immediately following correspondence with Sereno Watson of the Gray Herbarium at Harvard, who invited the teenager to serve as his assistant and to go to school at Harvard. There Fernald obtained his B.S. summa cum laude in 1897. He was already an assistant professor (1895) and bothered with no more degrees. By 1915 he was Fisher Professor of Natural History, Asa Gray's former position; in 1928, editor of the journal, *Rhodora*, and premier systematist of the eastern United States; in 1937, director of the Gray Herbarium. So to the end of his life.

Fernald's marriage was followed with a produce of two daughters and a son, about whom he possibly thought from time to time. He was short, somewhat rotund, mostly bald, and aristocratically short-bearded. He allegedly had a congenital heart condition, which however, seemingly did not limit extensive and vigorous field work in southeastern Canada during the first part of his life. Subsequently he compromised by working in tidewater Virginia where there were no steep mountains to climb. In later life, his vision was handicapped by cataracts.

Adjectives applied to Fernald include tireless, dynamic, enthusiastic, and vigorous in expression. In the throes of composition he loved to share. He read his manuscripts aloud to any and all within shouting distance. Some of his students heard his fermenting papers and review diatribes so many times that they could recite them back.

Fernald's two major productions were the eighth edition of *Gray's Manual* and his so-called nunatuk theory. Of the former, he had collaborated with B.L. Robinson in the preparation of the seventh edition of the *Manual* (1908). But the book was somewhat dated even when published—someone called it a cut-and-paste job. Fernald evidently decided then and there that it was going to be done right the next time. And for the next forty-two years, most of his efforts and voluminous publications were directed towards this goal. He barely made it. The manual, eighth edition, approximately 1,600 pages, appeared in print barely before his death in 1950. To the extent that we have "standard" manuals for the northeastern United States, these remain the Fernald and the **Gleason**-Cronquist (1963, second edition, 1991). Of the two, the latter is easier to use and now with the second edition more up to date, the Fernald remains more fully informative.

And historical plant geography, the Fernaldian nunatuks. The glaciers had repeatedly come across the upper half of the continent wiping out everything before them. If so, where did our present flora of the northern parts come from? Fernald provided an answer as early as 1925 (*Persistence of Plants in Unglaciated Areas of Boreal America*) derived from his work in southeastern Canada and the northeastern United States. Not everything was destroyed, he said; there were unglaciated areas, nunatuks, that were refugia for elements of the preexisting vegetation. After final retreat of the glaciers, these were the germinal sources of revegetation of the temperate parts of the continent. Fernald continued to pound on this and related theses concerning the history of the flora of the eastern United States and adjacent Canada and purportedly intended to put them together in unified form after completion of the manual. He ran out of time. Nonetheless, his efforts were enough to stimulate, perhaps even overstimulate, the thinking and the writing of an entire generation of plant taxonomists. I say overstimulate because glaciers seemingly become a central focus of explanation for any problem concerning the distribution of the plants one was studying. Manipulate the glaciers with skill or designate nunatuk refugia where necessary, and you could provide a satisfactory explanation. It was the Pleistocene era in North American phytogeography and taxonomy.

With A.C. Kinsey, Fernald was author of a book on edible wild plants. He prepared several generic monographs, for example, of *Potamogeton* and *Draba*. He was editor of *Rhodora* for thirty-two years and its chief contributor. There appeared many of his dissertations about glaciers and historical plant distribution. There also he published the results of his studies preliminary to the manual, revisions of anything "critical" among seed plants and ferns of his area, accompanied by descriptions of innumerable new varieties and forms, most of which were later taken up in the manual. Thumbing through successive volumes of *Rhodora* and the Harvard *Contributions from the Gray Herbarium,* one cannot but be awed by the hundreds of pages of Fernald's hypercritical writing based in large part on continuing field work, especially in Newfoundland, the Gaspé, and later the Virginia Coastal Plain. Certainly his knowledge of the flora of temperate eastern North America exceeded that of anyone then or since his time. One only adds the qualification that he perhaps too often overextended himself in describing as taxonomic varieties phenotypes that did not quite match the usual or idealized kind. This being so, one can be content that he dealt with varieties and forms, not new species like E.L. Greene or P.A. Rydberg. The fact that he has been blindly followed by many subsequent writers suggests the weight of his authority.

Rhodora was indeed an entertaining journal in these days, and it was Fernald who made it so. His lengthy taxonomic papers were often prefaced by an informal diary or travelogue of the trip that made the wonderful taxonomic and geographic finds possible. Three sentences from one of them: "Frank meekly asked, 'Are we going through to the Beach without a stop? I'm dying for a smoke.'" (*Rhodora* 43:511, 1941). And "We were promptly taken in charge by a plain clothes detective and held for some hours, while questioned by one officer and another as German spies who had been under observation for days." (*Rhodora* 43:516, 1941). And "It is not comfortable to dig thistles and fold tall specimens of them in the dark" (*Rhodora* 39:344, 1937). Can you imagine such appearing in one of today's "rigorously reviewed" journals, dull grey in format and content by comparison?

Then there were the Olympian Fernald reviews of the work of others, a type of commentary that made would-be authors tremble in their field boots. It has been said the Fernald presence served as a conscience for the quality of taxonomic publication in eastern North America; for example, E.D. Merrill observed "His trenchant criticism . . . [assisted] in maintaining the standards of American botanical scholarship." Presumably incompetent or slovenly authors were frightened away. But not George Neville Jones who published the first edition of his *Flora of Illinois* in 1945. The first printing of any book of state flora commonly has some defects. Under the title of *An Incomplete Flora of Illinois,* Fernald used 15 pages in *Rhodora* (47:204-219, 1946) to reprimand the author. Jones, unlike others, did not suffer such in repose. For in 1947, he thundered a response to shake the pavement of Divinity Avenue (location of the Gray Herbarium), one extending some 10 pages in the *American Midland Naturalist.* Those who enjoyed the show from the sidelines dubbed him "the man who dared to twist the lion's tail."

But we change after continued exposure to life. The lion in youth was able to breath poetical thoughts. For there has recently been discovered in the library archives at the University of Maine a multiparted poem or verse, *A Dream of the Woods,* reasonably attributable to the young Fernald (remember he grew up in Orono, Maine). See Mehrhoff (1990) for explanation and printing of the entire verse, the first four lines of which are as follows:

Merritt Fernald

> As seated in my city study drear,
> I linger o'er my volumes, lost in dreams,
> Sweet recollections now to me appear.
> And I am again following woodland streams.

(The above sounds like Fernald had been reading Poe.)

The later Fernald, a veteran of a thousand woods, and a million words perhaps was different. Possibly he indeed felt that the highest form of human culture existed only at Harvard and was improperly noisy about it. But his prepotent role in American systematics cannot be denied. Blessings upon him.

SELECTED BIBLIOGRAPHY

Ewan, J. 1971. M.L. Fernald. In: Dictionary of Scientific Biography. Vol. 4, pp. 583-584, with citations.
Fogg, J.M. Jr. 1951. Fernald as a teacher. Rhodora 53:39-43.
Mehrhoff, L.J. 1990. A dream of the woods. A recently discovered piece of early Fernaldiana. Rhodora 92:113-119.
Pease, A.S. 1951. Merritt Lyndon Fernald. Rhodora 53:33-39.

Fredrick Edward Clements (1874–1945)

Following **Warming,** the Danish enunciator of ecology, **Cowles** in the United States got in the first licks with his sand dune successions. It was Clements, however, who became and stayed as Mr. Ecology perhaps to the middle of the twentieth century. Then the walls of Jericho came tumbling down—as we will see.

Born in Lincoln, Nebraska, Clements stayed there the first thirty-three years of his life. And with good reason—after getting his undergraduate degree at the University of Nebraska, there was no better place to continue his training. These were the glory days of Nebraska. The magnetic **Bessey** had drawn together a group of evolving botanical stars such as Clements, Rydberg, and Roscoe Pound (but the latter later became a lawyer and dean of the Harvard law school). They called themselves the Botany Seminar. And there were ancillaries like the young novelist Willa Cather and the economist Alvin Johnson. So Clements continued for his Ph.D. (1898) and then as a member of the faculty. While there he married one of his students (1899), who finished her Ph.D. in 1904. Edith Clements was not the usual invisible botanical wife; she and her husband were a team until the end of his life. Not only was she a publishing co-worker, but she also became nurse and caretaker when Clements was afflicted with diabetes in the days before there

Frederick Clements

was insulin. The story that he would dictate reports to her while driving is not apocryphal (Ewan 1971). Wisely there were no children.

In 1907 the Clementses left Nebraska, he to become head of botany at the University of Minnesota. They stayed ten years. Then Frederick Clements left academia, going to the Carnegie Institute of Washington where and when most of his (their) major work was done, retiring in 1941. During these years with the institute, the home base was variable depending on the season, including Tucson, Arizona; Santa Barbara, California; and an alpine laboratory on Pikes Peak.

Among friends and amiable colleagues, Clements was evidently a nice comfortable chap. Although probably limited in small talk, he was quick and eloquent when speaking as an authority. Others outside his circle are said to have often found him aloof and cold, becoming arrogant if one dared to disagree with him. Roscoe Pound has said he was deeply religious; if so this was independent of the conventional church formalities. It is also related that Clements was a wide reader and interested in music and human problems.

Clements' first interest in botany was mycology. As early as 1894, he published a book on the genera of fungi (revised in 1931 with C.L. Shear), and other mycological publications followed. Pound and Clements together prepared the *Phytogeography of Nebraska* (1898). Clements wrote a book, *Plant Physiology and Ecology* (1907), but a caustic review by Barnes at Chicago possibly discouraged him from further efforts in physiology. But there were other pastures: *Rocky Mountain Flowers* (with Edith Clements), *The Phylogenetic Method in Taxonomy* (with his Carnegie colleague, Harvey Hall), and *Flower Families and Ancestors* (with Edith Clements). A neo-Lamarkian, he conducted a series of transplant experiments between 5,500 feet to the apex of Pikes Peak at 14,000 feet. He asserted that there was genetic alteration as a consequence of altitudinal climatic differences. It took another Carnegie team headed by Jens Clausen to clarify the interpretational smog.

Clements' supreme reputation, of course, is derived from the structuring of plant ecology, down to the last brick and mortar placement. I can only list some of his more significant publications, books, and journal papers. These began with his *Development and Structure of Vegetation* (1904) and *Research Methods in Ecology* (1905). (The latter was praised in the United States but said to be trivial by one of the German physiological ecologists.) Perhaps the most important of his major works was *Plant Succession: An Analysis of the Development of Vegetation* (1916). There he enunciated the word, and most subsequent publication was just that of adjusting the pieces. His thesis, with the inclusion of animals, was presented jointly with Shelford in *Bio-ecology* (1939). This book and the Weaver and Clements' *Plant Ecology* (1929) served as texts to pass the word to the unwashed (like Isely, a student at that time) who had not seen the more technical Carnegie *Plant Succession*. Then there were subsequent important journal papers like *The Relict Methods in Dynamic Ecology* (1934) and *The Nature and Structure of the Climax* (1936). The human welfare aspect was added in *Plant Succession and Human Problems* (1935) and *Climatic Cycles and Human Population on the Great Plains* (1938). Here Clements became a consultant to the U.S. Department of Agriculture in range management, forestry, and dust bowl rehabilitation.

But the preceeding does not explain the nature of the Clements doctrine. Let us try. Bare soil or sand is invaded by pioneer plants. In due time they change the demanding habitat to the extent that replacement with a different floristic population inevitably follows, the new being a second stage (sere) of habitat maturation. Successive seres then ensue until an equilibrium dictated by the climate is reached. This is the climax.

The climax formation, said Clements, is not an abstraction, but an organic entity ("The formation arises, grows, matures and dies."). Adapted to a given climatic region, it is an inevitable product of that climate. Clements described fourteen climaxes for North America. These he called formations. Subordinate units of the formation characterized by dominant or codominant species were named associations.

But variation in local conditions, for example, soil factors, may terminate succession short of the climax. Clements carefully defined perturbations and took care of them with such terms as preclimax, subclimax, and postclimax.

Frederick Clements

The Clementian concepts were as the holy Catholic Church, every detail preordained, defined, and carefully located in its appropriate place. This Clements enunciated as if his eyes were firmly fixed on the Deity perched upon the mantelpiece above him.

There was some carping about all of this, especially the view that the vegetational formation constituted a super-organism; many students perhaps more easily would have accepted an analogy. And the unitary definition of the climax exclusively on the basis of climate seemed a bit arbitrary. **Tansley,** Clements' British equivalent, thought the system too rigid, but he maintained a continued admiration of Clements' conceptual breadth. The major heretic was the American ecologist, **Gleason.** Gleason maintained that there was no such thing as a plant community except in a coincidental sense and that vegetational structure per se has no objective reality. The frequency or infrequency of the same species in the same "community," he said, was dependent on their ability to thrive in similar habitats and monitored by their distributional history and pure chance. He regarded the climatic climax as an aberration of Clements' mental ruminations. But Gleason was a Martin Luther in the Vatican Garden. Ecologists predominantly were sufficiently in love with the orderly beauty and functional usefulness of the Clementian edifice that Gleason's cries of anguish wrought but little.

Let's move up some fifty years, through a couple of generations of ecologists. Perhaps they feel it is sad to destroy a citadel, but if its foundations rest in sand, come down it must. They have pulled the Clementian structure asunder and stomped on the pieces. This has been possible, I think, because Clements was primarily an observer and philosopher rather than an experimental ecologist. Yes, he paid due obeisance to "the experiment," (e.g., his transplant studies). But somehow he seemingly assumed a priori that the climax exists and that all else necessarily must be assembled in this light. In striving for conceptual doctrine, he then depended primarily on his formidable intuition. This was not sufficient when put to critical analysis.

The philosopher Kuhn has said that scientists work within the constraints of a given paradigm. When a better one arrives, they shift gears and reorient the precepts of their philosophy and objectives of research. True, Clements had fallen of his own weight, but has a new paradigm arrived?

It then seems that Clements is a brave warrior of the past. Not at all. Shelford's *Ecology of North America* (1963) essentially uses the Clementian classification for major hierarchies, fitting in the animals at the subset level. Forest service and range management specialists still follow Clementian classification. Obviously, they remain valuable for convenient classification of North American plant community groupings, even if the taxonomy, as of **Linnaeus,** may be regarded as artificial.

Tansley often did not agree with Clements. But he said that he was "by far the greatest individual creator of the modern science of vegetation" (Tansley 1947).

SELECTED BIBLIOGRAPHY

Ewan, J. 1971. F.E. Clements. In: Dictionary of Scientific Biography. Vol. 3, pp. 317-318, with citations.
Schantz, H.L. 1945. Frederick Edward Clements. Ecology 26:317-319.
Tansley, A.G. 1947. Frederick Edward Clements. J. Ecology 34:194-196.

Agnes Robertson Arber (1879–1960)

Agnes Arber was (and is) an admired paleobotanist, botanical historian, comparative morphologist, botanical artist, and scientific philosopher, and she burgeoned at a time when professional opportunities for female scientists were mostly nil and negative.

Agnes Robertson's father was an artist; so was her sister, and, although obscured by her other attainments, so was she. She attended University College, London (1897-1898), and Newnham College, Cambridge, and received a D.Sc., University of London, in 1905. She derived inspiration from F.W. Oliver (pioneer paleobotanist) and served as an assistant to a sister-in-arms, amateur cytologist Ethel Sargent, who apparently had money and maintained a private laboratory called Jodrell Junior (senior presumably being the Jodrell laboratory at Kew). Ms. Robertson married E.A.N. Arber, a paleobotanist, in 1909. He died in 1918. They had one daughter.

Arber used a laboratory at Newnham College until 1927. Then she moved her work to a converted bedroom in her home, the site of activity the rest of her life. She had no professional income and no research support except for some small grants.

I have nothing regarding Arber's nonprofessional personality beyond reiteration among her contemporaries that she was kindly and helpful. That her fiscal means were limited is supported by statements that she lived frugally.

Arber was a loner. Her criteria for research were and are at odds with current holy doctrine. She decried team investigations in which "research becomes a form of routine activity. . . . One requires a generous measure of solitude rather than a laboratory where people are incessantly running in and out. . . . Meditation is essential to understanding," and "Independence is the essence of research" (Stearn 1960. p. 263). These specifications fitted her personality and would be appropriate for many investigators today.

So? Arber wrote eighty-four journal papers in addition to about a dozen books. A partial enumeration of the latter will give a notion of the woman and her works. Following an early skirmish with ancient plants (paleobotany), she turned to ancient books, the herbals, and their archaic woodcuts. She had not the means to trail about the major European libraries hunting superannuated illuminated manuscripts. So she limited herself to the printed herbals (primarily 1470-1670), many of which were conveniently available in the library of the British Museum. *The Herbals, Their Origin and Evolution* was her first book (1912). A rewritten second edition appeared in 1938. A recent (1986) third edition is a reprint of the second annotated by the classicist William T. Stearn of the British Museum. "The Arber" has remained the basic reference for the herbals despite much literature of the past half century, including Frank Anderson's spritely *An Illustrated History of the Herbals* (1977). Then came *Water Plants: A study of Aquatic Angiosperms* (1920; reprint 1963). And *Monocotyledons: a Morphological Study,* which developed the thesis, first suggested by **A.P. de Candolle,** that the monocot leaf is derived from a dicot petiole. These three persuasive books, magnificently illustrated by the author, are in no way compilations; they are syntheses strained through the meshes of the author's critical and contemplative mind. Their role as documents of continuing significance and usefulness is evidenced by the recent reprinting of all of them.

In subsequent years, her chutzpah, the search for the truth, proliferated into precarious terrain. How and why do we know what we think we know? What is the meaning of it all? One here enters the portals of a mix of science, metaphysics, and philosophy evidently stimulated by Spinoza and Goethe. For example, *The Natural Philosophy of Plant Form* (1950), *The Mind and the Eye* (1954), and *The*

Manifold and the One (1957). That Arber had not entirely left botany for the darkling universe of epistemology is, however, indicated by her last publication (1961, posthumous), *Theoretical Basis of Plant Morphology* in the (Peter Gray) *Encyclopedia of the Biological Sciences*.

Tansley, British ecologist, in 1952 said, "Dr. Agnes Arber is the most distinguished as well as the most erudite British plant morphologist" (Stearn 1960, p. 263). Amid other quotable laudatory ovations at the time of her death, I prefer the mild remark of Stearn's (1988, p. xxxi), "Continued esteem and use of her publications remain her best reward." A goal for any botanist.

SELECTED BIBLIOGRAPHY

Goodwin, H. 1970. Arber, Agnes Robertson. In: Dictionary of Scientific Biography. Vol. 1, pp. 205-206, with citations.
Stearn, W.T. 1960. Mrs. Agnes Arber: botanist and philosopher. Taxon 9:261-263.
Stearn, W.T. 1988. Introduction (pp. xxv-xxxii) In: D. Agnes Arber Herbals. Cambridge University Press, Cambridge.
Thomas, H.H. 1960. Dr. Agnes Arber, F.R.S. Nature 186:847-848.

Henry Allan Gleason (1882-1975)

In a generation really not too long departed, Gleason was of the illustrious in plant ecology, taxonomy, and plant geography. I suppose he now remains best known to the general botanical profession either or both as the man who dared to challenge **Clements** and as author of the three-volume *Illustrated Flora of the Northeastern United States* (1952). The latter in itself was enough to engage one for a lifetime. Seemingly for the author, it was mostly but a postlude.

Gleason was born in Illinois and went to school at the University of Illinois. There, quickly a botanical fanatic and gifted with a computer memory, he knew much of the midwest flora before finishing his master's degree. He received his doctorate at Columbia University (1910) in systematics, working under that diminutive tiger, **N.L. Britton,** director of the New York Botanical Garden. Thereafter he was successively at the Universities of Ohio and Michigan, not as a systematist but instead an ecologist, teaching, building ecological facilities, conducting research, and promulgating ideas out of tune with the times. In 1919, he accepted a position at the New York Botanical Garden and with it, a professional reversion back to systematics, except that he continued to talk and publish ecological heresy for some eight more years. Why the change? Gleason answers in a manuscript autobiography (1944): the "monotony of teaching those pharmacy students" and

Henry Gleason

"the desire for more time for research." Perhaps this is not the whole story, but in any event, he ceased being a professional peripatetic and spent the rest of his career, some thirty-one years, at the New York Botanical Garden. There he held several administration positions including head curator of the herbarium, assistant director and acting director of the garden, as well as being continuously active in taxonomic research. After retirement, he went back to ecology-plant geography, producing with Cronquist the *Natural Geography of Plants.*

What sort of a person was Gleason and what did he do, if anything, besides botany? Asking the same question for all of these essays and often finding but a dry well, I here have too many riches. These include especially a copy of an unpublished 50-page autobiography of Gleason's (1944) and a published appreciation by Cronquist (1976). Even so I had to hunt to determine whether he was a bachelor or not—he wasn't. He had three children.

Physically Gleason was a small, perhaps unprepossessing, bald-headed, well-mustached man with a somewhat weathered salamander-like appearance. He was one of quiet dignity and spartan nature who never uttered a word of profanity, never drank (but I have been told he was routinely attached to a curve-stemmed pipe), and who abominated "off-color" jokes.

The autobiography outlines Gleason's career with serious poker-faced satisfaction. It touches on details such as how much he was paid in successive appointments and describes successive operational catastrophes he inherited. For example, "The facilities were miserable my first five years . . . the students were of still lower caliber," and "Instead of a successful biological station, I was now in charge of thirty acres of Canada Thistles. . . . I went to work."

Cronquist portrays Gleason as one who was serious, yes, and confident of his ability, but who also had a continuous "puckish" sense of humor. For example, Gleason established a fictitious Society for the Preservation of Bumble Bees. On being asked about dues, he responded, "That's where you get stung." He smiled when told about the poison ivy growing about the garden, presumably believed by the unwary to be some handsome ornamental. He played bridge Sunday afternoons and kept cumulative scores for years.

Thus, Gleason's career included ecology, plant geography, taxonomy, and administration. His ecology was quickly innovative and iconoclastic, while his taxonomy contrasted in being conventional and almost ultraconservative (e.g., he hated to change plant names, and there are apocryphal tales of his hiding nomenclatural discoveries). In taxonomy, he left the theorizing to others, while in ecology, he was almost the only idea person who really competed with Clements.

Now about Gleason, the ecologist. He initially tested some of **Cowles'** generalizations about forest succession in southern Illinois. They did not work.

Perhaps stimulated by this frustration, and in order to present the relative commonness of species beyond statements such as "rare, abundant, scattered," etc., he used quadrants—counted numbers of individuals of each species in 10-foot square grids—and broke them into groups according to number. From this he developed a frequency index, said by others to be the first introduction of "statistical methods in ecology." Next came a Frequency Formula asserted to be "free from bias." Seeking mathematical justification, he developed an equation ultimately "proved" (so he says) by his son "in five minutes." Obviously one has to read the papers to understand what this is all about.

Gleason also worked with forest succession and its relationship to the original prairie in Illinois. There was no single climax forest type, rather there were at least three distinct ones. The once more widely distributed forests were now mostly seen in river valleys less exposed to prairie fires than the uplands. But prior to human (Indian) habitation, the prairie had been more cosmopolitan, occupying areas where it now was only represented by relicts. These investigations were widely acclaimed (Gleason says) as pioneering studies in historical phytogeography.

Came the atmometer, developed by Livingston, to measure evaporation, from which transpiration rate could be inferred. "The view was advanced that changes in the rate of evaporation were potent causes of succession, and various papers, based on field data, were written to prove it" (Gleason 1944). It seemed to Gleason, however, that this was backassward (but he would not have used this expression); rather he held that "it was the initiation of a succession that changed evaporation." So he did some work, pre-

Henry Gleason

sumably establishing his view, and published it. No one, he says, subsequently challenged his position.

A large area has more species than a small one. Is there a mathematical relationship between size and number? Yes, says Gleason. The increase in species number is logarithmetric; if so plotted, it leaves a straight line and, when tested, proved to "work." Because others said he was nuts, he made more tests on various vegetational types. The Gleasonian logarithmic approach was verified at last to his satisfaction.

And finally the association, as presented by Clements, now "elevated with a halo of mystery," a super-organism in which each individual has its individual function. But no! The association has no ordained structure, saith Gleason; it is chance, a patchwork, a coincidence for which there are no season tickets. Gleason presented these views at the meeting of the International Botanical Congress in 1926. They produced, he says, a major "commotion," seemingly equivalent to that of desecrating the American flag. Gleason evidently gave up on the issue. He was able, however, to draw satisfaction from the fact that people gradually came to him in subsequent years and said they agreed with him.

This then was essentially Gleason's last major effort in ecology. Thereafter, he was almost entirely a systematist.

All right, Gleason, the taxonomist. His pathway is easier to trace than his zigzag ecological career and contrasts oddly with it. For in ecology, he was an original and often iconoclastic idea person. In taxonomy, contrariwise, he was conventional and almost super-conservative. I don't think a single proposition about doing things differently came from him. He was a doer who left the theorizing to others.

Really, Gleason's taxonomic career started with his doctorate dissertation, a revision of North American *Vernonia* (Compositae) at the New York Botanical Garden in 1907. A backwash of that study was that Gleason hated that family thereafter and years later asked Cronquist to "do it" for the *Illustrated Flora*.

On returning to the New York Botanical Garden in 1919, he initiated a new South American program, and this was his primary research focus for nearly twenty years of traveling, collecting, and writing. Much of South America was yet so little explored botanically that every collection brought in numerous new species—Gleason described more than 1,000. The enormous Melastomaceae with innumerable species, yet but poorly known, became his speciality. There he wrote up some 500 new species and numerous genera. He wrote revisions of several major genera.

In 1939, Gleason switched back to the humdrum United States and started on a new edition of the *Britton and **Brown** Illustrated Flora*. I suppose the reason to be that **Fernald** at Harvard had been feverishly working on a new edition of the *Gray's Manual* for about thirty years, and the garden needed to stay in competition with Harvard. In any case, the *Britton and Brown*, last revised in 1913, and using the defunct American Code of Nomenclature, badly needed rewriting. Gleason farmed out a few groups to specialists, the major one being the Composites by Cronquist. Otherwise he did the whole thing, three volumes, complete rewriting, new illustrations, in a period of thirteen years—an absolute miracle of accomplishment!

While this was going on, one must keep in mind that Gleason also had major administrative positions at the garden as listed earlier. Also, with Edmund Fulling, he founded the *Botanical Review*, presently known to botanists whatever their speciality. He was a major force in the establishment of the American Society of Plant Taxonomists and drew up its administrative organization, which has stood the test of time.

Gleason's manuscript autobiography was written in 1943 (dated February 1944) before his formal retirement in 1950. The last paragraph ends, "Now I am again a botanical farmer, patiently steering my plow back and forth through the same soil that my ancestors tilled, trying to turn a straight furrow, and wondering if I shall reach the end of the field before winter sets in."

Well, he did beat the winter as to the *Illustrated Flora*. In retirement epilogue, however, he took on a book, *The Natural Geography of Plants*. At the age of about ninety, he had to decide he was not up to it anymore. The terminal chapters were written by Cronquist.

Gleason's career was distinctly bifurcate. His systematic efforts immediately brought praise and recognition. Acceptance of his role as the man who dared to bite the lion's (Clements') tail came only with the degeneration of the organismic climax.

SELECTED BIBLIOGRAPHY

Cronquist, A. 1976. Dr. H.A. Gleason, an appreciation. Garden Journal, New York Bot. Garden 26(2):56-59.
Gleason, H.H. 1944. Autobiography. 51 pp.
    Unpublished, reference copy at the New York Botanical Garden.
McIntosh, R.P. 1975. H.A. Gleason—"individualistic ecologist" 1882-1895. Bull Torrey Bot. Club 102:253-273.
Smith, A.C. 1951. Dr. H.A. Gleason. Garden Journal, New York Bot. Garden 1(1):53, 56, 64.

By permission of Barbara Yuncker, print courtesy of Hunt Institute for Botanical Documentation, Carnegie Mellon University, Pittsburgh, Pa.

Winona Hazel Welch (1896–1991)

Winona Welch was a bryologist; her subjects were mosses. She followed Truman Yuncker at DePauw University, Indiana, as the botanist of name. Both evidently were supreme teachers and both were systematists who published vigorously.

There has not yet been time for literature about Welch to accumulate except for the cited Festschrift from which this account is largely drawn. The Festschrift is a Hallelujah Chorus honoring Welch. One must allow for bias: the singers chosen are those who are unlikely to get off key.

The Festschrift includes a detailed "Biographical Essay" about Welch by Jeanne Goode, followed by a tabulation of education, professional positions, and a listing of her publications. Then there are individual tributes by (my count) twenty-six people, mostly about the personality, work habits, teaching, and research accomplishments of Dr. Welch.

Welch grew up on a farm in Jasper County, northwest Indiana, and began her education in a one-room school. She soon was intent on going to college. Sadly, support by her family was lacking—her father thought she should get married and "settle down." And college seemed not financially possible. The determined Winona, however, by doing secondary school teaching saved up $1,000 in four years. She entered DePauw University in 1919. There, initially, she had the notion of majoring in chemistry but was turned away. Then the magnetic Truman Yuncker, department head, brought her to botany. She graduated from DePauw (B.A. 1923), went to the University of Illinois (M. A. 1925), and finally to Indiana University (Ph.D., ecology, 1928). That was her formal education.

It now looked like Welch was a comer, but she yet had to choose the direction to take. The origin of her continued affair with Bryophytes (mosses and liverworts) came in part from Yuncker while she was still an undergraduate. Maybe the mosses were attractive for research because nobody seemed to know much about them. Charlie Deam, at this time the floristic magistrate of Indiana, facetiously assured her that there were only two kinds of mosses, big and small ones. He encouraged her to jump in and prove him wrong. She did. Returning to DePauw University as assistant professor in 1930, she took off in teaching and research. She was associate professor 1934-1939 and professor 1939-1961.

The initial years fairly well provided the pattern for the rest of Welch's personal and professional life. Work, work, work. Get up at 5:00 A.M. before the clamor of classes starts. Work over the weekend. Don't take vacations. A workaholic and then some. This was necessary because Welch wanted to do research and publish. But she had a horrendous teaching load; the usual topics of elementary biology plus such subjects as plant pathology, landscape gardening, and bacteriology. I don't know how many courses she routinely taught at one time. Apparently, if one can believe the chorus, she was a born teacher, a great teacher, the "best teacher I ever had." She swims in complimentary adjectives. But even though one is gifted, a heavy teaching load uses much time and competes with the other half of the pie, research; she neglected neither.

Welch participated in the doings of scientific societies to which she belonged, especially the American Bryological (and Lichenological) Society of which she was secretary treasurer for ten years. And she was active in several churches. Service! Those who devote themselves to service are rarely either rich or famous.

After Truman Yuncker's retirement, Dr. Welch inherited the headship of the Botany and Bacteriology Department that she held 1956-1961. She retired from teaching in 1961 and entirely in 1968.

But she stayed on, devoting herself to moss research and taking care of the herbarium. An honorary Doctorate of Science came to her from DePauw University in 1982.

What was the product of the research? In 1957 it was the *Mosses of Indiana,* a user-friendly and otherwise excellent state flora. And then the Fontinalaceae, believed by some to be the most poorly known and abjuratory of all moss groups. For example, Crum (1983) said that *Fontinalis* was surely the most difficult of mosses. In 1934 came the "Fontinalaceae of North America" in Grout's *Moss Flora of North America.* Then ultimately a monographic *Fontinalaceae* for the world, published after a ten-year wait (NSF finally came to the author's rescue, 1960). I have looked at this bulky monograph, but am not qualified to judge the systematics. The most obvious feature is that Welch seems to drown in the bureaucracy of collection citations of all material seen, given in the text and then essentially repeated in the index.

Welch started work on the Hookeriaceae and published it in *North American Flora* (1976). Eye failure probably terminated much work beyond this time.

Beyond the Jeanne Goode essay, in the cited Holmgren, Holmgren, and Buck (1988), tributes to Welch from twenty-six individuals were published, ranging from a short paragraph to about four pages. They had in common the deification of the subject. The following are a sampling of notes or quotes from this part of the Festschrift, plus some commentation by Isely.

L. E. Anderson: "Welch worked several summers at the H. S. Conard sponsored 'moss clinics' at the Iowa Lakeside Lab and one or more summers at the laboratory of A. J. Grout." These were important in that they gave Welch a chance to exchange views with other bryologists and avoid the parocialism of working entirely in isolation, as she was at DePauw.

K. W. Davis: "I saw her as a rebel. She chose to be single . . . didn't she feel society dictates? A delightful and hep woman."

R. F. Dawson: Compared Welch with her predecessor at De Pauw, Yuncker. "He was the dominant figure, the droll extrovert. She was the sensitive introvert."

The Yunckers: Welch had said that vacations were used for research. Well almost. She spent several summers at the Yuncker cottage and fished with Truman Yuncker. And "She wrote daily to her parents as long as they lived." That is probably stretching it.

W. C. Steere, of the next generation of bryologists and ultimately, president of the New York Botanical Garden: "Welch was unskilled in grantsmanship." Lacking funding for foreign field work she went on cut-priced commercial tours. There was always some unguided leisure time. She then collected mosses.

Beyond the statement that Welch was a phenomenon at DePauw, Steere made the observation that in effect, her treatments were mechanical, and lacked depth, i.e. did not consider phylogeny; there was no interpretation or opinion expressed of the interelationships of species and genera.

The *Moss Flora of Indiana* lacks such speculation. But one does not expect it in a floristic manual. Phylogeny might presently be incumbent in the monographic *Fontinalaceae,* but Welch's writing was at a time when it was less obligatory than now (1994), to consider (or speculate) about evolutionary relationships.

Welsh was curator of the herbarium and, although she did no research on lichens, established the lichen collection as part of the herbarium. These herbaria and the general DePauw herbarium are now at the New York Botanical Garden.

Winona Welch missed much that most of us consider desirable, if not necessary, for a balanced life. If this be a deficiency, she more than made it up in the totality of her accomplishments and her extraordinary talent for passing enthusiasm to others. She is one of the parents of present day bryology.

SELECTED BIBLIOGRAPHY

Crum, H. 1983. Mosses of the Great Lake Forest, 3d ed. University of Michigan Herbarium, Ann Arbor, Mich.

Holmgren, P.K., N.H. Holmgren, and W.R. Buck (Sponsors). 1988. Winona H. Welsh Festschrift. Brittonia 40:113-171.

Welch, W. H. 1957. Mosses of Indiana. Bookwalter Company, Indianapolis. 478 pp.

———. 1960. A Monograph of the Fontinalaceae. Martinus Nijhoff, The Hague. 357 pp.

Winona Welch, who lived into the 1990s, is botanist number 101 and brings us to the troubled present in our roll call of plant scientists. Glad we got slightly acquainted. Of course, there have been many botanical souls other than these. But we probably won't have an opportunity to patrol the scrolls of time for botanists 102, 103, etc. "Good-Bye Mr. Chips."

Index of Botanists

In the text botanists are given in chronological order so that some concept of the evolution of the discipline may be provided. The alphabetic index here tells where you may find a botanist you are interested in.

Adanson, M., 97
Arber, A.R., 331
Aristotle, 3
Bailey, L.H., 274
Banks, J., 110
Bartram, J., 80
Bartram, W., 80
Bary, H.A. de, 213
Bauhin, G., 49
Bentham, G., 163
Berkeley, M.J., 170
Bessey, C.E., 237
Boch, H., 23
Bower, F.O., 264
Brefeld, J.O., 220
Britton, N.L., 280
Brongniart, A.-T., 167
Brown, R., 132
Brunfels, O., 17
Buffon, G.-L, 82
Burbank, L., 247
Camerarius, R.J., 74
Candolle, Alp. de, 178
Candolle, A.P. de, 145
Carver, G.W., 285
Cesalpino, A., 39
Chapman, A., 181
Chase, M.A., 303
Clements, F.E., 326
Clusius, 43
Cordus, V., 29
Coulter, J.M., 251
Cowles, H.C., 306
Darwin, C.R., 184

Deam, C.C., 293
Dioscorides, 10
Dixon, H.H., 308
Dixon, H.N., 291
Dodoens, R., 32
Dutrochet, R.J.-H., 139
Eaton, A., 143
Eichler, A.W., 223
Engelmann, G., 188
Engler, H.G.A., 230
Fernald, M.L., 321
Fries, E.M., 154
Gerard, J., 46
Gesner, K., 35
Ghini, L., 20
Gleason, H.A., 334
Goebel, K.R.E. von, 267
Gray, A., 191
Grew, N., 68
Haberlandt, G., 258
Hales, S., 77
Haller, V.A. von, 94
Hedwig, J., 101
Henslow, J.S., 157
Hildegard, 14
Hitchcock, A.S., 296
Hofmeister, W.F.B., 210
Hooke, R., 65
Hooker, J.D., 196
Hooker, W.J., 148
Ingenhousz, J., 104
Jones, M.E., 254
Jussieu, A.-L. de, 118
Knight, T.A., 121

Lamarck, J.B., 113
Leeuwenhoek, A., 60
Linnaeus, C., 86
MacDougal, D.T., 299
Mattioli, P.G., 26
Mendel, J.G., 203
Mohl, H. von, 176
Moldenhawer, J.J.P., 127
Nägeli, C.W. von, 200
Nuttall, T., 151
Persoon, C.H., 124
Pfeffer, W.F.P., 241
Priestley, J., 107
Pringsheim, N., 207
Pursh, F.T., 136
Ray, J., 56
Sachs, J. von, 216
Saussure, N.T. de, 129
Schimper, A.F.W., 271
Schleiden, J.M., 173
Scott, D.H., 261
Small, J.K., 311
Strasburger, E.A., 233
Tansley, A.G., 315
Thaxter, R., 278
Theophrastus, 6
Torrey, J., 160
Tournefort, J.P. de, 71
Van Helmont, J., 53
Van Tieghem, P.E.L., 225
Vries, H. de, 244
Warming, J.E.B., 227
Welch, W.H., 340
Willstätter, R., 318

Incidental Index

Adanson, validation of (Jussieu), 118-20
Air contributes to plant nutrition? (Van Helmont), 53-55
Alternation of generations (Hofmeister), 210-212
American botany, domination of for 40 years (Gray), 191-195
American Code of Nomenclature (Britton, Small), 282, 283, 313
Antithetic vs. homologous theories (Pringsheim), 208, 209
Apostles for botany (Bailey, Bessey, Coulter), 237-240, 251-253, 274-277
Aristotelian scholasticism (Cesalpino), 40-42

Bartram's Travels (W. Bartram), 81
Beitrage zur Anatomie der Pflanzen (Moldenhawer), 127
Bibliotheca Universalis (Gesner), 36
Bildungsgewebe and Dauergewebe (von Nägeli), 201
Binomial nomenclature (Linnaeus), 90
Biological latin (Linnaeus), 90
Bluthendiagramme (Eichler), 223, 224
Botanical exploration (Nuttall, J.D. Hooker), 151, 152, 196, 197
  financier of (Banks), 110, 111
Botanical Features of North American Deserts (MacDougal), 301
Botanicorum facile princeps (Brown), 135
Botany and poetry (Bailey), 274
Botany of Darwin, flower pollination adaptations (Darwin), 186

Botany textbooks (Gray, Bessey), 193, 239
The British Islands and Their Vegetation (Tansley), 316
Brownian movement (Brown), 132
Bryology (H.N. Dixon, Welch), 291, 292, 340-343
  pioneer of (Hedwig), 101-103

Catastrophism, Cuvier (Lamarck), 116
Causae et cures (Hildegard), 15
Causes (Aristotle), 5
Cell theory (Dutrochet, Grew, Hofmeister, von Mohl, Moldenhawer, von Nägeli, Pringsheim, Schleiden), 3, 14, 127, 128, 173-177, 201, 208
Cellular morphogenesis (Strasburger), 233-235
Chlorophylls (Willstätter), 319
Classes for children (Henslow), 158
Classification, flowering plants (Bessey), 239, 240
Classification that worked (Linnaeus), 86-92
Climax formation, an organic entity (Clements), 328, 330
Cohesion theory (H.H. Dixon), 309
Commentarii...Dioscorides, 14 editions (Mattioli), 26, 27
Comparative anatomy (Van Tieghem), 225
Cordus manuscripts, publication of (Gesner), 36

347

INCIDENTAL INDEX

Cruÿdeboek (Doedens), 33
Curator of experiments (Hooke), 65

Darwin
  introduction to botany (Henslow), 157, 158
  spokesman for (Gray), 194
Das Pflanzenfamilien (Engler), 230, 231
De Plantis Libri XVI (Cesalpino), 40
Description of plant world, last at species level (A.P. de Candolle), 145, 146
Die Pflanzen und ihr Leben (Schleiden), 174
Diffusionism (Alp. de Candolle), 179

Ecologists (Tansley, Gleason), 316, 336-337
Ecology (Warming), 227
  of deserts (MacDougal), 300-301
Élan vital (Dutrochet), 140
Epigenesis vs. preformation and other controversies (Haller), 94-95
Evolution, a chain of life (Lamarck), 115, 116
Expansive generalizations (Buffon), 83, 84

Father of botany? (Theophrastus), 6-9
  German "fathers" of botany (Brunfels, Boch), 17, 23
Father of natural classification? (Linnaeus), 86-89
Ferns and flowering plants (W.J. Hooker), 149-150
First esteemed botany teacher (Ghini), 20-21
First native American botanist (J. Bartram), 80, 81
Fixity of species (Brongniart, Linnaeus, Lamarck), 86-91, 114
Flora Americae Septentrionalis (Pursh), 137
Flora of Indiana (Deam), 294, 295
Flora of North America (Gray, Torrey), 161, 162, 192, 193
Flora of the Southeastern United States, 1903 and 1933 (Small) 311-313

Flora of the Southern United States (Chapman), 181-183
Flower, fruit, seed, importance in classification (Gesner), 37
Flowering plant families, natural classification of (Jussieu), 119
Flower parts in classification (Cordus), 29-30
Flower pollination adaptations (Darwin), 186
Fossil plants (Brongniart), 167, 168
Founder of descriptive mycology? (Clusius), 44
Fungi
  in culture (Brefeld), 221
  distribution of (Berkeley), 171
  natural classification of lichens (Fries), 154

General Historie of Plants (Gerard), 47
Genera of North American Plants and a Catalogue of the Species of the Year 1817 (Nuttall), 152
Genera Plantarum (Bentham), 166
Genera Plantarum (Jussieu), 119
Genetics and Darwinian evolution (Mendel), 203-205
Genus man (Tournefort), 71-72
Geotropism (Knight), 122
Geschichte der Botanik (Sachs), 217
God had appointed him (Carver), 290
Grass person (Chase, Hitchcock), 296-298, 303-305
Gray's *Manual of Botany of the Northern United States,* five editions (Gray), 192
Gray's *Manual,* 8th ed. (Fernald), 321
Green plants purify air only in light (Igenhousz), 105
Grew and Malpighi: contemporaneous, but independent (Grew), 69
Grundzuge der Wissenschaftlichen Botanik (Schleiden), 173

Herbalists, herbals (Boch, Dioscorides), 10, 23
Herbal reprinted in twentieth century (Gerard), 46

INCIDENTAL INDEX

Histoire naturelle des animaux sans vertèbres (Lamarck), 115
Historia Plantarum (Ray), 57
Historia Stirpium Indigenarum Helvetiae (Haller), 95
Hortus siccus (herbarium) (Ghini), 21
How water gets up a tall tree (H.H. Dixon), 308, 309

Illustrated Flora (Britton), 280
Illustrated Flora of the Northeastern United States (Gleason), 334, 335
Index Herbariorum (Hitchcock), 298
Index Kewensis (J.D. Hooker), 198
  financed by Darwin (Darwin), 187
Inheritance of acquired characters (Lamarck), 113, 116
Institutiones rei herbariae (Tournefort), 71, 72
Introduction to Systematic and Physiological Botany (Nuttall), 152

Journal of Ecology (Tansley), 316

Kew Gardens, directors, (J.D. Hooker, W.J. Hooker), 149, 197
Kreuterbuch (Boch), 23

Land flora, origin of (Bower), 264
Lehrbuch der Botanik (Sachs, Strasburger), 217, 235
Letters to Royal Society (Leeuwenhoek), 61
Life histories (Pringsheim), 208, 209
Lindley's, *Natural System of Botany* (1830) (Torrey), 161
Linnaean system (Eaton), 144
Logical division of knowledge (Aristotle), 4, 5
Lois de la nomenclature botanique (Candolle, Alp. de), 179
Lyceum (Aristotle, Theophrastus), 3, 6

Magnification lenses, and new observations (Leeuwenhoek), 60-62
Manual of Botany for the Northern States (Eaton), 144
Materia Medica, longevity of (Dioscorides), 10, 12-13
Methodology, supreme precision of (Thaxter), 279
Micrographia (Hooke), 66
Microscopic cross section of cork (Hooke), 65
Microscopist (von Mohl), 176
Missouri Botanical Garden (Englemann), 189, 190
Mosses of the world (Dixon, H.N.), 291, 292
Mutation man (de Vries), 244, 245
Mycological classification (Fries), 154, 155
Mycologists (Persoon, Thaxter), 125, 126, 278, 279
Mycology (de Bary, Berkeley, Brefeld), 170, 171, 213, 214
  descriptive, founder of? (Clusius), 44

Native and cultivated plants (Clusius), 44, 45
Natural history, all phases (Theophrastus), 8, 9
Natural philosopher (Lamarck), 114-116
New botany (Sachs, Strasburger), 216, 234, 235
New York Botanical Garden (Britton), 280, 281
  administrator (Gleason), 335, 338
"Now with our friendly labours..." (Gerard), 47
Nunatuk thesis (Fernald), 322-323

One-man agricultural extension service (Henslow), 158
Opera Botanica (Gesner), 36
Organizer of the efforts of others (Engler), 230-232
Organographie der Pflanzen (Goebel), 268, 269
Organs, phylogenetic abstractions (Goebel), 268, 269
Origin of cultivated plants (Alp. de Candolle), 179
Osmosis (Dutrochet), 139, 140

Osmotic pressure (Pfeffer), 242

Paleobotany (Scott), 262
Peanuts and sweet potatoes (Carver), 287
Pflanzenphysiologie (Pfeffer), 242
Philosophy of conservation (Tansley), 316
Phylogenetic speculation (Goebel), 368, 369
Physica (Hildegard), 15, 16
Physiological anatomist (Haberlandt), 259
Physiologist/horticulturist (Knight), 121, 122
Pinax Theatri Botanici (Bauhin), 50
Pioneer plant breeder (Burbank), 247-249
Plant anatomy (Grew), 68-70
Plant chemist (Saussure), 129, 130
Plant description, diagnostic (Cordus), 29, 30
Plant ecology, structure of (Clements), 327-329
Plantesamfund (Oecology of Plants) (Warming), 228, 229
Plant Geography on a Physiological Basis (Schimper), 271, 272
Plant kingdom, classification of (Scott), 263, 263
Plant metabolism (Saussure), 129, 130
Plant nutrition (Van Helmont), 53-55
Plant physiologist, pioneer (Hales), 77, 78
Plant physiology (MacDougal, Pfeffer, Sachs), 217, 218, 241, 243, 299-301
Plant inheritance experiments (Mendel), 203-205
Plant recognition, taught verbally (Cordus), 30
Plant respiration and photosynthesis (Priestley), 108, 109
Plants
  and medicine (Dioscorides, Hildegard), 10-13, 15-16
  and poetry (Doedens, Bailey), 33, 274
  sex in (Camerarius), 74, 75
  and air purification (Igenhousz, Priestley), 105, 108
Plant succession leading to equilibrium (Cowles), 307
Plant tissues (Moldenhawer), 127, 128
Polynomial nomenclature (Bauhin), 50
Popular botanical books (*The Green Leaf*) (MacDougal), 302
Potato blight (de Bary), 214
Première British ecologist (Tansley), 316
Prince of systematic botanists (Bentham), 163-166
Professor, University of Glascow (W.J. Hooker), 149
Progenitor of "natural classification" (Ray), 58

Reduction division (Strasburger), 235
Research criteria (Arber), 332
Respiration and photosynthesis (Saussure), 130

Sap movement in plants (Hales, Knight), 78, 122
Science and metaphysics (Arber, Buffon), 83, 332, 333
Sexual reproduction in the lower plants (Hedwig), 102
Sexual role of flowers, in Linnaean classification (Camerarius), 75, 76
Sixty years of botany (Bower), 265, 266
From special case to generalization (Brown), 134
Species Muscorum Frondosorum (Hedwig), 102
Species Plantarum (Linnaeus), 87-92
Spontaneous generation (de Bary), 213, 214
  denial of (Leeuwenhoek), 63, 213

INCIDENTAL INDEX   351

Synopsis Fungorum (Persoon), 125
Synopsis Plantarum (Persoon), 125
Systema Naturae, 12 editions
  (Linnaeus), 86, 90
Systematics
  North American pioneer
    (Englemann), 188-190
  revisionary (Bentham), 164, 165
Systematist (Gleason), 337, 338

The versatile Hooker (J.D. Hooker),
  197, 198
Tissue culture (Haberlandt), 259
Traité de Botanique (Van Tiegham),
  226

University of Chicago, botanical
  glory days at (Coulter), 252

Vegetable Staticks (Hales), 78
Vegetation, dynamics and classification of (Cowles), 306, 307

Water transport, mechanical (H.H.
  Dixon), 308, 309
Weed wagon (Deam), 294
Weiditz, Hans (Brunfels), 17, 18
Willow tree experiment (Van
  Helmont), 54
Woman suffragist (Chase), 304, 305